Pure and Wholesome

A collection of papers on
water and waste treatment
at the turn of the century

Papers selected by the Committee on History and Heritage
of the American Society of Civil Engineers

AMERICAN SOCIETY OF CIVIL ENGINEERS FOUNDED 1852

Published by the
American Society of Civil Engineers
345 East 47th Street
New York, New York 10017

cover photo

New Reservoir, 1889, for Wheeling, West Virginia. The contract for the work was let in 1888 to W. M. Chesholm, East Liverpool, Ohio. The 3,000,000 gallon reservoir was 50 by 665 feet in plan and 11½ feet deep. It was built for $30,000. This photograph, taken in 1917, illustrates the effect of using unfiltered river water. It was necessary to clean the silt out of the reservoir annually! Rapid sand filters were installed in 1925 to provide pure water for Wheeling.

Library of Congress Catalog Card No. 81-70989
ISBN 0-87262-290-8
Manufactured in the United States of America.

PURE & WHOLESOME

A Collection of Papers on

Water & Waste Treatment at the Turn of the Century

FOREWORD

This publication is one of a series on the history of civil engineering sponsored and published by the American Society of Civil Engineers through its Committee on the History and Heritage of American Civil Engineering. Using the rich mine of historical information buried in the early Transactions of the Society a series of monographs have been planned. This is the 5th in the series.

To many the problems of the degradation of the environment are of recent origin, but in fact the history of sanitary engineering (later subsumed under the title environmental engineering) clearly demonstrates a major commitment on the part of engineers toward an improved urban environment. The turn of the century represents the nascent period for modern environmental engineering in terms of the application of scientific knowledge to the problem of water and waste under treatment and the organization of the profession along modern lines. Thus, rather than a collection of papers from different periods in the history of environmental engineering, this seminal period of two decades is represented by four papers from leading engineers on critical problems in the field. Through these papers and the published discussions one gains a clear insight into the ferment in the field regarding such topics as filtration and something of the deep concern of engineers to improve the environment. In order to place these papers in historic context a brief history of the subject, prepared for the ASCE Centennial in 1952, is included as the introductory paper.

This monograph was prepared under the aegis of the Committee on the History and Heritage of American Civil Engineering of the ASCE. The members of this Committee are:

Messrs.: Neal FitzSimons, Chairman
John A. Adams, Jr.
David Caplan
H. Hobart Holly
Emory L. Kemp
Alan L. Prasuhn

CONTENTS

PURE & WHOLESOME

The city is a symbol of civilization, first established as a place to worship the gods, then its development is a center of culture, trade and government. Indeed, the history of mankind could well be interpreted through a study of the leading cities in history from ancient Babylon to Victorian London (1). Even Utopian writers have cast their visions of a perfect society in terms of an ideal city. The modern city with its extensive network of transportation, communication and service systems is a product of the Victorian Era. The city, regardless of time or place, reflects in a striking way the values of the society which built it.

Asa Briggs, a leading historian of 19th century Britain, with unusual insight has written:

> The building of the cities was a characteristic Victorian achievement, impressive in scale but limited in vision, creating new opportunities but also providing massive new problems. Perhaps their outstanding feature was hidden from public view - their hidden network of pipes and drains and sewers, one of the biggest technical and social achievements of the age, a sanitary 'system' more comprehensive than the transport system. Yet their surface world was fragmented, intricate, cluttered, eclectic and noisy, the unplanned product of a private enterprise economy developing within an older traditional society(2)

In reading contemporary accounts of Victorian technology one is swept up in the romance and adventure of the engineer as a hero constructing monumental works on an unprecedented scale. This conjures up visions of Brunel and the building of the Great Western Railway and the launching of the Great Eastern, the Roeblings and the Brooklyn Bridge, Captain Eads and the Mississippi Bridge at

Saint Louis or the construction of the Trans Continental Railroad, immortalized by Hollywood. If, however, the modern city is the greatest social and technical achievement of the 19th century then a closer look at the "network of pipes, drains and sewers" and those who built them is long overdue. In a very real sense this is an unheralded achievement of enormous magnitude and importance.

From surviving Roman acqueducts to E. S. Chesbrough's famous Gothic Revival water tower in Chicago (1867) the concern of engineers to supply water to cities was a matter of quantity and hence primarily a problem of hydraulics. If sanitary (i.e. environmental) engineering is perceived in these terms then the field has an ancient lineage. If, however, sanitary engineering is concerned with water quality and the provision of copious amounts of water that is "pure and wholesome", to use a Victorian phrase, then the advent of the field could be dated as 1807 when sand filters were first used in the Glasgow water system. The prototype slow sand filter was first used in the Chelsea Water Works Co. in London in 1827 and found its first application in America in Richmond in 1832. These early examples of water filtration were intended to provide clear and potable water for both industrial uses, particularly in textile works, and for drinking water.

During the 1840s and for the next two decades environmentalists in England rallied around the "Sanitary Idea." In the best British tradition the movement was led by enlightened amateurs, most notably Sir Edwin Chadwick. Cleanliness was not only considered next to godliness, but was seen as a means of improving public health through the elimination of "noxious vapours" and filth diseases, principally typhoid fever and cholera.

The celebrated Broad Street Pump episode in 1854 firmly established the contaminated water from this pump as the source of a cholera epidemic. The next year the municipal authorities in London required all water companies to supply filtered water. An understanding of how diseases are transmitted in water was not known at the time and only emerged after Pasteur's pioneering work on the germ theory of disease, which was announced in 1857. By the 1870s Koch et al. had verified this theory and it had found general acceptance. This is an outstanding example of a scientific idea giving rise to a series of technological innovations, since the term theory spawned a series of investigations which led directly to modern methods of filtration for both water and waste water, the use of chlorine as a disinfectant and the establishment of bacteriological techniques for measuring water quality.

Until 1887 American engineers borrowed extensively from British practice and to a lesser extent from work in Germany and did not contribute significantly to the advancement of the field. In that year the Commonwealth of Massachusetts established the Lawrence Experiment Station for the study of water quality. Hiram F. Mills (1836-1921) director, recruited an outstanding multi-disciplinary staff which made noticeable contributions to the problems of water treatment, including establishing the efficacy of filtration in removing typhoid germs. Equally important, members of this team became leaders in the field and were responsible for establishing the practice of sanitary engineering as we know it today. At this time water, sewerage and solid waste systems were increasingly coming under municipal control, whereas, the design of sanitary engineering works was in the hands of private consulting firms. Members of the Lawrence fraternity were leaders in both the public and private sectors.

From 1887 to 1893 experiments were made with chlorine as a disinfectant for drinking water and sewage, but the first continuous use of chlorine was not begun in America until 1908. It was the end of this first decade of our century that saw the introduction of the trickling filter and the Imhoff tank for sewage treatment.

Thus, the early development period of water treatment and sewage and solid waste disposal methods occurred between the establishment of the Lawrence Experiment Station in 1887 and the introduction of the Imhoff tank in 1910. Although a definitive history of environmental engineering remains to be written it appears that this period of just twenty-three years may be the most important in the history of the subject. As a result, all but one of the papers selected for this monograph are from this period. In these papers and associated discussions leading engineers discuss the most appropriate means of providing water of acceptable quality and the problems of pollution from sewage and solid wastes. The emphasis is on filtration, with fascinating discussions of filtration versus watershed protection to insure an uncontaminated source of water. Also the question of the necessity of treating sewage if in fact filtration is effective in producing a water that is "pure and wholesome" produced a lively debate. Attitudes and practices concerned with environmental quality were developed in this formative period which persist to this day.

The introductory paper was prepared for the Centennial of the A.S.C.E. in 1952 and provides an historical context in which to place the other papers.

We see from these papers that at a time of rapid urbanization which resulted in unprecedented conditions of crowding and sheer numbers of people, the health standards in American cities were not only maintained but greatly improved not by spectacular developments in medicine but by the provisions of copious amounts of safe drinking water to millions of city dwellers. It is one of the little known triumphs of engineering.

REFERENCES

1 Toynbee, Arnold, ed. Cities of Destiny
pub. McGraw Hill Book Co.
N. Y. 1967

2 Briggs, Asa, Victorian Cities
pub. Harper Colothon Book Co.
N. Y. 1963

AMERICAN SOCIETY OF CIVIL ENGINEERS
Founded November 5, 1852
TRANSACTIONS

Paper No. 2608

PROGRESS IN SANITARY ENGINEERING IN THE UNITED STATES

By E. Sherman Chase,[1] M. ASCE

Synopsis

Sanitary engineering—a science dealing with water supply and purification, sewerage and sewage treatment, drainage, refuse collection and disposal, and insect and vermin control—has achieved a remarkable progress record in the 100 years since Louis Pasteur initiated his experiments in bacteriology. Insect-borne and water-borne diseases have been conquered through the persistent efforts of early practitioners of sanitary engineering. This paper reviews the development of sanitary engineering in the United States. Prominent milestones are listed and related to the pattern of growth of the science. Future prospects are briefly discussed.

Introduction

Sanitary engineering deals with those factors of environmental sanitation which profoundly affect man's well-being. The factors generally accepted as within the province of sanitary engineering are water supply, water purification, sewerage and sewage treatment, drainage, refuse collection and disposal, insect and vermin control, and, to a limited extent, the sanitation of foods.

In dealing with each of these factors the engineer must be guided not only by the laws of mechanics, hydraulics, and the other physical sciences, but by the laws of chemistry and biology. Until the so-called "germ theory of disease" was put forth, the civil engineer who designed works for the acquisition and distribution of water and the disposal of the by-products of man's activity was concerned mainly with physical laws. Furthermore, until the health hazards of intestinal organisms and the transmission of yellow fever, bubonic plague, and other infectious bacteria by insects were recognized, the application of engineering skills to environmental factors was without thought of benefit or harm to human health.

Historical Background

Development of Bacteriology.—Almost coincident with the founding of the American Society of Civil Engineers, a series of events was taking place in France

[1] Partner, Metcalf & Eddy, Boston, Mass.

which in subsequent years stimulated the development of that branch of civil engineering now called sanitary engineering. These events were the researches of Louis Pasteur, first in the diseases of wine and beer, then in the diseases of silkworms and of fowl, and finally in the diseases of man himself.

Many years were to pass before the discoveries of Pasteur, Robert Koch, Karl J. Eberth, and other scientists working in the embryonic field of bacteriology, were accepted by the medical profession or utilized by the engineer. Nevertheless, it is not too great an exaggeration to claim that the sanitary engineer's origin, albeit feeble, dates to the time of the founding of the ASCE and was coincident with the work of Pasteur and contemporary bacteriologists and chemists.

Conquest of Infectious Diseases.—Before proceeding with a review of the advance in sanitary engineering during the past 100 years, it is well to consider briefly the plagues which sanitary engineering has abolished, or nearly abolished, from western civilization. For example, cholera, local to India prior to 1817, spread throughout the world during the rest of the nineteenth century. Some 8,500 deaths in Hamburg, Germany, as recently as 1892, were attributable to this disease.

John Snow, in London, England, demonstrated in the famous Broad Street well case that cholera infection was carried by water polluted with excrement from an existing case of the disease. This demonstration had little effect on American practices inasmuch as the sanitary watchword of the period was "cleanse," and as late as 1885 the reduction of incidence of "filth diseases" was attributed to cleanliness. Eventually, of course, the method of transmission of cholera and the proper preventive measures were generally recognized, and the last reported death from this disease in the United States occurred in 1911.

Yellow fever, the dreaded Yellow Jack of Cuba, Panama, and the Gulf coast, swept up the Mississippi Valley from New Orleans, La., in the summer of 1870, with 100,000 victims and 20,000 fatalities in its wake. In *Harper's Weekly* of that year there appeared these words relating to those tragic days:

> "This fragrant beauty seems the mark of Death,
> The whispering south wind is his poisoned breath;
> We weary for these warm bright days to end,
> The summer lingers, at what fearful cost.
> O, pitying God, in mercy to us send
> The white gift of Thy frost!"

Those were the days before Walter Reed and his fellow workers demonstrated the role of the mosquito in the spread of yellow fever, but "the poisonous breath of the south wind" is an apt term for those infected insects. However, even close to the end of the nineteenth century, evil emanations and mysterious miasmas from the earth were accepted in many quarters as the causative factor for yellow fever, malaria, and even typhoid fever. The Johns Hopkins Hospital, in Baltimore, Md., which opened in 1889, had a basement insulated with a thick layer of asphalt, to check dangerous exhalations from the soil.

There were 1,282 deaths from yellow fever in Havana, Cuba, as late as 1896; but by 1902 this scourge was erased thanks to the labors of William C. Gorgas and his corps of doctors and engineers. Nevertheless, yellow fever appeared again in New Orleans as late as 1905. The construction of the Panama Canal was made possible by the eradication of the yellow-fever-bearing mosquito in the Canal Zone. In this connection, it has been stated that Ferdinand de Lesseps, during his abortive attempts to build the canal, lost one third of his white workers by yellow fever.

The history of the prevalence of that other mosquito-borne disease, malaria, and its control, more or less parallels that of yellow fever, although the eradication of malaria has not been nearly so complete. Proper drainage and the elimination of breeding places played an important part in the control of mosquitos and their spread of yellow fever and malaria.

It is to the elimination of typhoid fever that the sanitary engineer points with pride. George C. Whipple, M. ASCE, wrote in 1908 regarding typhoid:

> "The indications are that the disease is soon to become much less prevalent than it is today. In another generation the death rate ought not to be over one-third or one-fourth of what it now is, and who can say that the disease will not be all but obliterated just as smallpox has almost vanished from our midst."

These were prophetic words. In the United States, the typhoid fever death rate, which approximated 35 per 100,000 population in 1900, dropped to less than 1 per 100,000 in 1950.

Sanitary Engineering in the United States

The development of sanitary engineering in the United States since 1852 can be roughly divided into quarter-century periods from the middle of the nineteenth century to the middle of the twentieth century. In reviewing this development through each of the four periods, it is not amiss to refer to some of the more outstanding concurrent historical events.

Period from 1850 to 1875.—The acquisition of the present area of the continental United States by discovery, annexation, cession, and purchase (subsequent to the Louisiana Purchase of 1803) ended in 1867 with the purchase of Alaska from the Russians. The discovery of gold in California in the late 1840's stimulated the development of the west coast at the beginning of the period, and westward migration followed.

Although the Civil War retarded the normal progress of the country, particularly in the South, land grants to Union soldiers following the war and federal assistance to the railroads accelerated the westward flow of population. In the matter of sanitation, the death toll from diseases and wounds on the battlefield and in the prison camps led to well-intentioned but, in the light of modern knowledge, uninformed attempts to ameliorate the conditions causing the spread of infection. In other sections of the world, the dreadful conditions in war hospitals during the Crimean War of the middle 1850's aroused the pity and ire of Florence Nightingale. Similarly, the conditions during the Franco-Prussian War of 1870 evoked in Pasteur the concept of antiseptic treatment of wounds.

In 1850, Lemuel Shattuck prepared a forward-looking report to the Massachusetts Legislature on a plan for "Promotion of Public and Personal Health" in which occurred this significant recommendation:

> "* * * that, in laying out new towns and villages, and in extending those already laid out, ample provision be made for a supply, in purity and abundance, of light, air, and water; for drainage and sewerage, for paving, and for cleanliness."

During this quarter century there developed the first glimmering of insight into the causes of those diseases now termed infectious, and the beginning of the recognition of the relationship of environmental conditions to the prevalence of those diseases. In the Second Annual Report of the Massachusetts State Board of Health, Dr. George Derby, its secretary, made the startling assertion that:

> "It may be stated in round numbers that one person out of every thousand in Massachusetts, between the ages of 5 and 70, dies yearly from typhoid fever."

Nowhere in Derby's report on his inquiry into the causes of typhoid does he indicate that in his opinion the public water supplies of Massachusetts were in any way related to the prevalence of the disease. However, the English had begun to connect the occurrence of typhoid with the pollution of water supplies, and a few physicians in the United States suspected such a relationship.

According to George W. Fuller, M. ASCE, there were eighty-three public water supplies in the United States in 1850. By 1870 the number had increased to only two hundred forty-three. Generally, these supplies were confined to the larger cities and were derived from nearby rivers or ponds. Although an unsuccessful attempt at water filtration had been made at Richmond, Va., in 1832, not until 1872 was the first successful slow sand filter in the United States constructed at Poughkeepsie, N. Y., in accordance with the designs of James P. Kirkwood, Past-President ASCE. Construction of the filter at Poughkeepsie was followed 2 years later by a similar one at Hudson, N. Y. These two filter plants represent the only successful American attempts to filter public water supplies during the third quarter of the nineteenth century.

Prior to 1850, sewerage and drainage were installed in very few communities. In some cases, the discharge of fecal matter into drains was actually prohibited. This was true in Baltimore, even as late as 1910. Not until about 1850 did the water-carriage system for liquid wastes begin to be adopted. The first sewerage system in the United States designed on engineering principles was that of Brooklyn, N. Y. This installation was based on plans prepared in 1857 by Julius W. Adams, Past-President and Hon. M. ASCE. In 1858, E. S. Chesbrough, Past-President ASCE, presented his first report on a sewerage system for Chicago, Ill., and in 1874 J. Herbert Shedd, M. ASCE, prepared designs for the Providence, R. I., system. These three installations appear to be the only instances of the application of engineering principles to sewerage design in the period from 1850 to 1875.

The few sanitary engineering projects cited represent just about the entire development of that branch of engineering up to 75 years ago. The succeeding quarter century, however, is marked by the real beginning of modern sanitary engineering.

Period from 1875 to 1900.—The last quarter of the nineteenth century was free from wars of major intensity, although there were relatively minor conflicts, such as the Russian-Turkish War, the conquest of Egypt by the British, the Japanese War with China, the Spanish-American War, and the Boer War. Compared to the Civil War in the United States and the Franco-Prussian War of the previous quarter century, these wars were relatively insignificant in their effect on the rest of the world. The Spanish-American War, however, with its dreadful toll from typhoid fever in the army camps, aroused the United States to the necessity for preventive measures against the spread of the disease.

In these last years of the nineteenth century, rapid development in many fields of human activity resulted in the increasing concentration of population in cities, the exploitation of natural resources, the production of many materials and mechanical devices, and, above all, a tremendous expansion in the engineering arts and sciences. The development of agricultural lands and mines in the western part of the United States, the production of more and more wealth, and the increasing demand for improved living conditions turned men's thoughts more intensively to the means for betterment of those environmental factors relating to comfort and health. In this period also the bacterial origin of certain diseases became generally accepted, and epidemics were no longer attributed to acts of God. This was also the era of the American pioneers in sanitary engineering.

During the middle and latter years of this quarter century there appeared a group of pioneering chemists, biologists, and engineers, whose fundamental investigations at the Lawrence Experiment Station in Massachusetts prepared the way for the application of scientific principles to water purification and sewage treatment. These were the days in which a group of brilliant young men, under the leadership of Hiram F. Mills, Hon. M. ASCE, began careers which have meant so much to the practice of sanitary engineering in the United States. Their accomplishments, like the shot at the Concord Bridge (in Massachusetts in 1775), were "heard around the world." This group consisted of the chemists Allen Hazen, M. ASCE, and Fuller, and the biologist and outstanding teacher, William T. Sedgwick of the Massachusetts Institute of Technology (M.I.T.) at Cambridge. Coworkers and pupils of these men, and pupils of their pupils, probably constitute the bulk of sanitary engineers in the United States today, and their influence has been felt far beyond the confines of North America.

A partial list of early leaders in the profession would include the following men: Rudolph Hering, M. ASCE; W. R. Nichols and T. M. Drown of M.I.T.; Leonard P. Kinnicutt of Worcester Polytechnic Institute at Worcester, Mass.; George E. Waring, Jr.; Whipple; Harry W. Clark; Stephen deM. Gage; Frederic P. Stearns, Past-President ASCE; Harrison P. Eddy, Past-President ASCE; F. Herbert Snow, M. ASCE; Frank A. Barbour, M. ASCE; Leonard Metcalf, M. ASCE; and Emil Kuichling, M. ASCE.

The first, although short-lived, course in sanitary engineering, was established at Columbia University at New York, N. Y., in 1886. The second course, established at M.I.T. in 1889, has continued to this day, although it became a graduate course a few years ago.

Naturally, the investigations that received the greatest stress in this period were those relating to the purification of water supplies. The design of the early slow sand filters at Poughkeepsie and Hudson was based on Kirkwood's studies of European filters, but not until the middle 1890's were slow sand filters, based on scientific principles demonstated at the Experiment Station, installed at Lawrence (Fig. 1). The spectacular drop in the typhoid incidence at Lawrence following the installation of these filters demonstrated beyond question the role of polluted water in the spread of the disease, and served as an object lesson that has never been forgotten.

Although slow sand filters were proving their worth and practicability with the relatively clear waters of New York and New England, they had proved unsatisfactory for treating the muddy waters of the West and South. Even before the

FIG. 1.—INTERMITTENT SAND FILTER, LAWRENCE, MASS., 1906

work at Lawrence, the demand for clean water by the paper industry led investigators and promoters to devise filters that could be cleaned by mechanical means and hence used with muddy waters. These attempts were only partly successful, and only after thorough and painstaking experiments conducted at Louisville, Ky., under Fuller, from 1895 to 1897, were the fundamental requirements for the design of the so-called American or mechanical filter established. Heretofore these filters were operated with little or no period for presettling, and without coagulation or with half-hearted attempts at coagulation. In New Orleans one filter company constructed a large plant that failed most ignominiously and at great financial loss to the promoters. It was not until the results of research carried on by Robert Spurr Weston, M. ASCE, were available that the method by which muddy Mississippi waters could be purified was demonstrated.

Between 1875 and the end of the nineteenth century, engineering studies and reports were made on the sewerage and drainage of numerous individual cities and towns, and several systems were actually installed. Early sewers were of vitrified clay or cement pipe, in small sizes, and of brick in large sizes. The first large sewers of concrete were built in Washington, D. C., in 1885.

Prior to 1900 few sewage treatment plants existed in the United States; more than 90% of the nation's sewage was discharged without treatment to streams, lakes, or the oceans. Those treatment plants that were in operation were intermittent sand filters (Fig. 2) or contact beds in the smaller communities, or chemical precipitation plants, such as those at Worcester and Providence. The former plant was the largest in the United States, and served a population of about 118,000.

Fig. 2.—Original Open Slow Sand Filters at Lawrence, Mass.

Period from 1900 to 1925.—The peaceful advent of the twentieth century was soon marred by the Japanese-Russian War, followed in due course by World War I. In spite of the retarding effect of war, the first quarter of the century marked the beginning of vast social, economic, and industrial changes in the United States, one of which—the widespread adoption of mass production—released men's physical and mental energies for the development of better and more healthful standards of living, in spite of continued concentration of population in great cities.

At the start of the period neither yellow fever nor typhoid fever had been conquered. Yellow fever was still a menace in the southern states, and typhoid was rampant in many northern and southern cities. Typhoid epidemics running to hundreds of cases were not uncommon.

As dark as the picture of water-supply sanitation was in general, Boston, Mass., New York, Providence, Rochester, N. Y., San Francisco, Calif., and other smaller cities, had sought out unpolluted upland supplies, thanks to the concept that "innocence is better than repentance." Unquestionably, however, these cities, and others with respectable water supplies, suffered from the sanitary sins of cities like Niagara Falls, N. Y., whose then unpurified water supply spread disease and death far and wide.

Strange, is it not, that a period at first free from the plague of war but afflicted with plagues of the flesh should before its end see the first great World War and the nearly complete eradication of many infectious diseases? Before the quarter century had ended, nearly every grossly polluted water supply in the United

FIG. 3.—EARLY CHLORINATION APPARATUS

States had been either purified or abandoned for a new and safe supply. By the end of the quarter century the great boon of chlorination had been demonstrated (Fig. 3) and applied to almost all surface sources, yellow fever had been eradicated, and malaria and other infectious diseases had been reduced. This progress was achieved partly as a result of better medical knowledge, but also in part as a result of the broad application of sanitary engineering to environmental factors. Credit for the reduction in deaths from infections is due to no single group, but should be shared by physicians, chemists, and biologists as well as by engineers.

In the field of sewage treatment shortly after the turn of the century the trickling filter was introduced from England—first at Reading, Pa. (in 1908) following experiments at Columbus, Ohio. The Imhoff tank, combining clarification and sludge digestion, came into favor about 1910, and soon became the most popular method of preparing sewage for secondary treatment processes. This popularity has continued to a lesser degree even to the present, particularly in the case of small installations. The disposal of sludge, still a problem, was usually accomplished by means of drying beds or sludge lagoons. Filter presses were used at chemical precipitation plants, and the vacuum filter was introduced at Milwaukee, Wis.

In this period the activated sludge process, inspired by research work at the Lawrence Experiment Station, but developed in England, began to receive serious attention in the United States. The first installation was at San Marcos, Tex. (in 1916), followed during the next 10 years by several others, that at Milwaukee (Fig. 4) being the most widely known. It was at Milwaukee that the pioneer development of a process for the manufacture of a salable product from the sludge was carried out.

FIG. 4.—JONES ISLAND ACTIVATED SLUDGE PLANT, MILWAUKEE, WIS.

Coincident with the increase of urbanization, there had developed problems incidental to the collection and disposal of solid wastes from towns and cities—such as garbage, rubbish, and street sweepings. During the period when certain diseases were attributed to the spontaneous generation of mysterious miasmas in decomposing organic matter and filth, efforts were made, more or less effectually—generally less—to cleanse cities of these solid wastes. Systematic private and public collections of the wastes began toward the end of the nineteenth century, and incinerators for rubbish and reduction plants for garbage were installed by a few cities before the beginning of the twentieth century.

During the first quarter of the twentieth century there was a time when the reduction of garbage became relatively popular, because of the recovery of grease and tankage. However, the introduction of mixed refuse incinerators by J. T. Fetherston, M. ASCE, at Staten Island, N. Y., was followed by the adoption of this method of refuse disposal by many other cities.

The demonstration of the part played by insects and rodents in the spread of disease led to greater efforts toward securing prompt removal and disposal of material serving to harbor the pests. The introduction of the motorcar and the gradual elimination of stables from urban areas did much to diminish the prevalence of the "filthy fly."

Sanitation of the production and handling of foods, particularly milk, was first given serious consideration early in the 1900's. Pasteurization of milk, a joint development of the chemist, the biologist, and the engineer, was recognized, before the end of the first quarter of the twentieth century as essential for the protection of milk quality. The decrease in the death rate among young children is undoubtedly due to a considerable degree to the widespread adoption of pasteurization.

This period marks the initial activity of the Cincinnati, Ohio, station of the United States Public Health Service, under the direction of Earle B. Phelps, in the study of stream pollution and water-supply quality requirements. There was also an increasing interest all over the United States in the prevention of stream pollution, although remedial measures were retarded by World War I.

During this same period the most spectacular victories were won by the sanitary engineer in the battle to stem the tides of death. In these years yellow fever was conquered, and the prophecy of Whipple regarding typhoid came to pass. There remained, however, many tasks related to general improvement of environmental conditions and to the refinement in techniques and skills. The pioneering period was over, and the day of consolidation of victories was at hand.

Period from 1925 to 1950.—The quarter century just ended has been a period of social and economic convulsions, first, the great economic depression, and then World War II and the atomic bomb. Nevertheless, progress in sanitary engineering and efforts to prevent the incidence and spread of disease have continued. As previously stated, however, the dramatic accomplishments of the sanitary engineer in the saving of life were consummated by about the middle 1920's and the period of refinement of design and particularly of the mechanization of sanitary works had begun.

During the second quarter of the twentieth century many modifications and variations of previously used water purification practices occurred. Among the developments of this period were: New methods of coagulation, changes in details of filter design, greater reliance on chlorination, new techniques in taste and odor control, and the greater use of filtration for improvement of chemical and physical qualities rather than for bacterial removal only.

In sewage treatment, however, there has been much more progress in the installation of treatment plants than had occurred in all preceding years. This resulted, first, from the adaptability of the activated sludge process for treatment of wastes in large cities and, second, from the development of high-rate modifications of both the activated sludge and the trickling filter methods of treatment. Other advances such as the digestion of sludge in separate treatment tanks, and the greater use of mechanical dewatering of digested sludge, were in the developmental stage barely a quarter of a century ago but have now become almost standard practice. All these advances have produced treatment plants of high mechanical complexity, but with far less space and labor requirements per unit capacity than the older plants.

During the second quarter of this century, more and more attention has been paid to the reduction of the pollution of waterways. Laws and regulations have been promulgated, enforcement agencies have expanded, public educational campaigns in behalf of clean waters have been waged, and hundreds of sewage treatment plants have been built. In spite of all these efforts, much remains to be done, particularly in regard to elimination of pollution by industrial wastes. This is the next broad field for the sanitary engineer to tackle, but he will need the help of the industrial engineer and the chemist as the problem is one of great complexity.

Conclusion

This brief discussion has necessarily omitted a multitude of interesting and significant details in order to present the progressive development and the glorious achievements of sanitary engineering during the past 100 years. Stress has been placed on the relation of sanitary engineering to the prevention of disease and to the promotion of health.

The Future.—What of the future? No man can say! The great days of conquest of disease in the United States by the sanitary engineer are in the past. From now on, he will hold the gains and press on to develop new and better techniques in his art. However, when the war drums no longer beat and the battle flags are furled, there are many other areas of the world where he can resume his struggle with those malignant environmental conditions that cause death and disease.

In the United States there will continue to be great demand for clean streams, lakes, and ocean shores. To restore waterways to the maximum degree of cleanliness consistent with economic limitations will require much research in the field of industrial wastes treatment and large expenditures on the part of industry in the application of such research.

In view of the need for conservation of natural resources it is not unlikely that in arid regions, reclamation of sewage and other liquid wastes may be realized. Much is to be done in this field. Salvage of the solid constituents in sewage and wastes will require more consideration, and in this connection research will undoubtedly provide answers to problems now unsolved. It is possible that, in the economical development of atomic energy, ways may be found to concentrate the soluble solids of waste liquors and thus permit their recovery or easy and cheap destruction. Atomic energy may also offer a method for economical distillation of fresh water from the ocean. In general, however, it is probable that the future, in so far as one can look ahead, will show more refinement in the basic techniques of sanitary engineering, the development of new techniques and continued research for better and cheaper methods in the purification of water, the treatment of sewage, the control of insects and vermin, the disposal of solid refuse, and the sanitation of foods.

Pasteur wrote many years ago, "It is within the power of man to cause all parasitic diseases to disappear from the world." Within his sphere and in the United States the sanitary engineer has almost reached this goal in less than 100 years of progress.

E. Sherman Chase (1884-1969)

A native of Massachusetts, E. Sherman Chase was a member of the MIT Class of 1906. Following work in Worcester, Mass., New York, and Reading, Pa., he served for six years as Assistant Sanitary Engineer on water supply investigations with the New York State Department of Health in Albany. He was made a partner in 1927 and continued in that capacity until his retirement in 1960. Later he retained his office with the firm and continued active in a consulting capacity and with his writing on technical subjects. He was a recognized authority here and abroad in the fields of water supply, and sewage and waste disposal. He held professional registrations in 13 states, and was called upon many times for expert testimony in court cases. He was deeply interested in matters relating to engineering ethics and responsibilities.

Mr. Chase was an active participant in the work of professional societies, serving in official capacities and contributing over one hundred technical papers. He served as president of the New England Water Works Association 1933-34; of the New England Sewage Works Association (now New England Water Pollution Control Federation) in 1953. He was a counselor of the American Institute of Consulting Engineers 1957-1960.

In 1960, the annual E. Sherman Chase Award of the New England Water Pollution Control Association was established in his honor. In 1968, the Institute of Water Engineers of Great Britain established the Friendship Award to be presented to a member of the American Water Works Association "who has been outstanding for his furtherance of friendly, helpful and professional relations between the water engineers of the United States and Canada and those of Great Britain." E. Sherman Chase received this award the first year it was presented; it was a fitting honor to a distinguished engineer and gentleman.

AMERICAN SOCIETY OF CIVIL ENGINEERS.
INSTITUTED 1852.

TRANSACTIONS.

Paper No. 1025.

ADDRESS
AT THE ANNUAL CONVENTION AT FRONTENAC, THOUSAND ISLANDS, N. Y., JUNE 26TH, 1906.

THE DEVELOPMENT OF WATER SUPPLIES AND WATER-SUPPLY ENGINEERING.

BY FREDERIC P. STEARNS, PRESIDENT, AM. SOC. C. E.

In selecting a topic for the address which the Constitution requires me to deliver on this occasion, I have found that there is ample precedent for speaking on the subject with which I am most familiar, namely, water supplies and water-supply engineering.

I was the more ready to choose this subject because it gives me an opportunity to speak from my personal experience of the advantages of co-operation between engineers and those engaged along other lines of applied science.

It would be futile in a short address to attempt to trace the beginning of water-supply engineering. That it was not in its infancy in the Roman era is a matter of common knowledge, both from the carefully written descriptions of the water-works of that era and from the remains now to be seen of the magnificent arches and other works near Rome and in former Roman provinces, which prove the great ability of the water-works engineers of that period.

Works for both public water supply and irrigation existed at a much earlier period in Greece, as is attested by Homer in the following description of the gardens of Alcinous:

"Two plenteous fountains the whole prospect crowned:
This through the gardens leads its streams around,
Visits each plant, and waters all the ground;
While that in pipes beneath the palace flows,
And thence its current on the town bestows;
To various use their various streams they bring,
The people one, and one supplies the king."*

Traces of water-works have been found near cities which flourished in earlier periods; and, as human nature and the human brain have changed but little since the beginning of written history, it can hardly be doubted that from the time when people first gathered together in large cities there has been some public method of providing the necessary water for drinking and other domestic purposes.

In the Middle Ages, and even in the period of the Renaissance, when the fine arts, literature and architecture flourished, the water-works engineer, if he existed, built few if any works worthy of especial notice.

Of what may be called modern water-works, London probably furnishes the earliest example. Before these were built there were, as is often the case, minor works for bringing water from springs near the city. One of them was completed in the year 1285, and others from time to time afterward, and there was obtained at a still later date a larger supply by pumping water with tidal power from the Thames at London Bridge.

The first extensive modern water supply, however, was that provided by the New River Water-Works of London, built by Sir Hugh Middleton, in the four years from 1609 to 1613. The water was obtained from springs near the River Lea, distant about 20 miles in a direct line from London, but the open canal through which the water flowed to the city made many windings to avoid hills and valleys, so that it had a total length of 38 miles. The canal had an inclination of 3 in. per mile; its width averaged 18 ft., and its depth seldom exceeded 5 ft. The springs failed to fur-

* Odyssey, Book VII.

nish the required quantity of water, and were supplemented by a direct connection with the river. These were very great works for those days, when, to quote words written in 1835, "the science and practice of civil engineering were very little known."

In 1633 London is said to have been well supplied with water from pipes in all streets, and in nearly every house the rent of which was from $75 to $100 per year.

The water companies had troubles then as well as now. The average remuneration for water was one farthing a barrel, but it is stated that "notwithstanding such palpable evidence of its cheapness, some persons have been loudly querulous concerning the high and unconscionable rate charged for water." Their complaints led to a parliamentary investigation.

It is also noted among the troubles of the period that persons living near the open aqueduct used it for bathing. Under the laws then in force they could be prosecuted only for trespassing, for which the penalty was transportation. Considerations of humanity are said to have prevented the water companies from prosecuting the offenders.

After London was well supplied with water, 163 years elapsed before a public water supply worthy of notice was provided for any of the larger cities of America. The first one to be supplied was Boston, into which water was introduced by a water company in 1796. The water was taken by gravity from Jamaica Pond, situated a short distance outside the city limits. Philadelphia built its own works, which were supplied with water by pumping from the Schuylkill River, and were first operated in 1801. New York was first furnished with a water supply by the Manhattan Company at about the same time. In all these cities the water was distributed through hollow logs, which are sometimes found well preserved at the present day.

The earlier water-works in America seem not to have been well appreciated, as the revenue derived from the Philadelphia water-works three years after the introduction of water was only $1 800 a year, and up to 1815 not enough had been realized from water rates in any one year to pay for the operation of the works.

The water supplies of New York and Boston continued to be wholly inadequate to the needs of those cities for many years, as money was not plenty, and the cities were loath to embark upon great enterprises with which they were unfamiliar. After long delay, however, they proceeded to build permanent works on a most liberal scale. New York, especially, displayed great courage in deciding to build an aqueduct 40 miles long, having a capacity of 80 000 000 gal. per day, from the Croton River to the city. This work involved the great difficulty of conveying the water over the Harlem River at a height of more than 100 ft. above its surface.

Having once begun the construction of modern water-works, the rapid growth of American cities and the lavish use and waste of water have required a very rapid development of water-works systems in all parts of the country, and works of unprecedented magnitude are now being undertaken with less hesitation than the much smaller works of many years ago. The most notable of these recent undertakings is that of the City of New York for bringing water a distance of 100 miles from the Catskill Mountains, at an estimated cost of $162 000 000.

Let us now leave the development of water-works as a whole, and consider the development of their various parts and the influence which water-works engineering has had upon engineering along other lines.

The Romans were masters of the art of building masonry aqueduct bridges and aqueducts, and the advance upon their practice up to the present time is not very striking. We now build, however, aqueducts of much greater size, and have a knowledge of hydraulics and of levelling which enables us to produce much more efficient works.

The Romans excavated long tunnels—one of them is said to have had a length of 3 miles—through which to convey water, but these must have been especially difficult to excavate before the days of explosives.

Their greatest trouble was that due to the absence of suitable pipes for conveying the water across low lands, and it was this difficulty which caused the construction of the aqueduct bridges to which reference has already been made.

They had lead pipes, made by folding sheet lead and soldering the edges, and used tile and stone pipes in some instances, but none of them compared in efficiency with the modern iron pipe.

The date when iron pipes were first used for water-works is uncertain; there are instances of their use as far back as the middle of the eighteenth century, but they did not come into general use until after the beginning of the nineteenth century.

In 1746 the Chelsea Water Company, of London, laid a 12-in. flanged pipe to a reservoir at Hyde Park, which was taken up and relaid in 1791 "in consequence of its joints being perished." Another pipe laid a few years later in the Kensington Gardens was taken up in 1819 on account of incrustations on the interior surface of the pipe. A 5-in. iron pipe was laid to supply water to Edinburgh in 1787. The London companies changed from wooden pipes, which were in general use, to cast-iron pipes, between the years 1810 and 1820.

Mr. Thomas Simpson, the engineer of the Chelsea Water-Works, of London, designed the first bell and spigot pipes, about the year 1785, and at that time laid a few lengths of this pipe with lead joints as an experiment.

Philadelphia in 1804 laid about three-fourths of a mile of iron pipe as an experiment, and in 1817 began to use cast-iron pipes with bells, similar to those now in use. These pipes were imported from England.

It is of interest to note that as early as 1810 a 15-in. flexible cast-iron main with ball and socket joints at intervals was laid across the River Clyde to supply water to a portion of Glasgow. A trench was first dredged in the bed of the river, then the pipes, attached to wooden frames, were connected on land on one side of the river, and subsequently dragged across it until one end reached the opposite shore, the operation being aided by pontoons, much in the same manner that such work is done at the present time. The credit for the design of this siphon is due to the celebrated James Watt, the inventor of the steam engine, who, for some years, practiced civil engineering.

"A patent Steam Engine" was erected by the New River Water Company, of London, in 1787, and it is possible, though not probable, that this was the first application of steam for pumping a city water supply.

The steam engines first used for pumping water for Philadelphia, in 1801, were unique machines, as the walking-beams, flywheels, connecting-rods, and even the cold-water pumps, were made of wood. As indicating the difficulty of constructing such machines in this country at that time, it is said that the main steam cylinder of one of the engines was cast in two parts, and that nearly four months were occupied in boring it out fit for use. The boilers were wooden boxes bolted and braced on the outside. "The fire-box inside the boiler was of wrought iron, with vertical flues of cast-iron."

There has been a very great advance in steam-engine practice since those early days, and the water-works pumping engine, which has been developed by some of our most able mechanical engineers, has, much of the time, been in advance of other steam engines in efficient performance. It is worthy of note, however, that for a long time some of the Cornish engines, used in the mines in Cornwall, were the most efficient machines then in use, this result being obtained by a system of co-operation and nearly continuous tests, the results of which were published monthly.

The triple-expansion water-works pumping engine of the present day is a very perfect and smooth-running machine, which, with a high degree of economy, may lift a sufficient quantity of water to provide for the requirements of 300 000 people.

In the art of dam building the engineers of water supply and of irrigation have been rivals for many years. In India, and in recent years in the arid lands of our own country, the irrigation engineers have been designing and constructing great dams of masonry and of earth. Up to the present time, however, the water-supply engineers have created the greater works.

The New Croton Dam, completed within the current year, with a height from the lowest point of its foundation to its top of 297 ft., is the greatest of masonry dams, although the mass of masonry is small in comparison with that of the Great Pyramid.

The San Leandro Dam of the Contra Costa Water Company, which supplies Oakland and other cities near San Francisco with water, is the highest of existing earth dams. It is an interesting feature that this dam, as well as three notable dams connected with the San Francisco Water-Works—two of earth and one of concrete—were shaken by the earthquake of April 18th, 1906, when under full water pressure, without any serious result.

NORTHEASTERLY PART OF SPOT POND RESERVOIR.

Although the art of dam building has made such progress that great dams like those just mentioned have been designed and built of sufficient strength to withstand safely the tests of practical use, the theory of the design of both masonry and earth dams is only imperfectly understood. The rules given in the textbooks for designing masonry dams do not take account of some important forces affecting their stability, most of which are the result of temperature changes. Several of the more recent failures of masonry dams, most of them on southern rivers, could without doubt be traced to the neglect of the effect of forces which are not sufficiently considered in making designs.

Experiments made during the last twenty years to determine the laws governing the filtration of water through sand and soils of various kinds have furnished much additional information for use in designing earth dams, and it has been found that, with proper designs, earth containing a large percentage of fine particles may be substituted successfully for a masonry or clay puddle core-wall.

There are some water-supply features which are not creditable to the American people. First, the great waste of water which prevails in most of our cities, and second, the taking of water directly from polluted rivers without either filtration or long storage.

There is, among those who govern our large cities, an unreasoning opposition to the introduction of water meters, which are conceded by nearly all who know about such matters to be the most efficient means for diminishing the waste of water. The recent action by the government of one of our larger cities emphasizes this point. This attitude seems to have some relation to the commonly expressed view that water should be as free as air, and that on sanitary grounds it is not well to restrict its use, but no one has yet demonstrated the sanitary advantages of a leaky faucet or a defective ball-cock.

Experience shows that meters prevent waste, but do not prevent the use of as much water as is required for sanitation. Waste not only adds greatly to the difficulty and expense of obtaining a sufficient water supply, and requires the expenditure of money which might well be applied to improving the quality of the water by filtration or otherwise, but it adds greatly to the difficulty and expense of the sewerage system, especially where pumping and purification of the sewage are necessary.

There is undoubtedly a rapidly growing opinion among water-works engineers and those connected with water-works management that the general application of water meters is desirable, and the public and the governing bodies of our cities will in time become educated so that they will accept this view.

While I have said that the taking of water directly from polluted rivers is not creditable to the American people, I should add that under the leadership of the engineers and sanitarians of the present day this discreditable feature is fast disappearing. There is also to be said, as an excuse for the practice years ago when many such water systems were inaugurated, that the purity of water was then judged mainly from a chemical standpoint. Bacteriology was unknown, and the efficacy of filtration was not appreciated.

To show the point of view held thirty-three years ago, I quote from a paper written by one of the most prominent water-works engineers of that time:

"The object of filtration is to remove the visible impurities of water and render it colorless. * * * * * There is also, to a limited extent, a lessening of the chemical impurity."

On the other hand, although the effect of filtration in rendering water wholesome was not so well known, there was ample precedent for filtering river waters, as it had been practiced in Great Britain since about the year 1811. In 1827 there were enough places where filter-beds were in use in England and Scotland to induce Mr. Thomas Simpson to make a trip of investigation, and he then visited filtration works in various places which were said to have proved completely efficient.

In 1835, a writer, after referring to the entire success of the filter-beds built by Mr. Simpson in 1829, offers this prophecy:

"Probably at no very distant period the desirable practice of filtering the whole of the water supplied from rivers to the inhabitants of great towns for domestic use may be universally adopted."

While this prophecy may have been fulfilled in England from the time when it was made, it is only approaching fulfillment in America.

DARK HOLLOW POND, ADJOINING SPOT POND RESERVOIR.

The first well-equipped plant in America for filtering water through sand was built at Poughkeepsie, N. Y., from plans prepared by the late James P. Kirkwood, Past-President, Am. Soc. C. E., and adopted by the city in 1869. This municipality had the honor of possessing the principal filtration plant of this kind in the United States for many years.

The work of the State Board of Health of Massachusetts, in regard to the purification of water and sewage, is well known, but I should leave out an important part of the development of water supplies and water-supply engineering, were I to omit it; moreover, the organization of this work gives me one of several opportunities to speak from my own experience of the advantages of co-operation between engineers and those engaged along other lines of applied science.

This Board, by a law passed just twenty years ago, was given general supervision of the streams, water supply and sewage disposal in Massachusetts, and was required to advise cities and towns as to the appropriateness of sources of water supply and the best method of disposing of the sewage.

Mr. Hiram F. Mills, the eminent hydraulic engineer, and Chairman of the Committee on Water Supply and Sewerage of the Board, after learning how little definite information was then available on many phases of these subjects, with wonderful foresight initiated the now classical examinations and experiments which have resulted in so great an increase in the knowledge possessed by the water-supply and sewerage engineers of the present day. In this work he was ably supported by the Chairman of the Board, Dr. H. P. Walcott, an eminent sanitarian. The speaker was at the time the Chief Engineer of the Board.

For making these examinations and experiments a staff of chemists, biologists and bacteriologists was engaged, comprising as leaders such men as the late Thomas M. Drown, Professor of Chemistry in the Massachusetts Institute of Technology, and afterward President of Lehigh University, and Professor William T. Sedgwick, of the Massachusetts Institute of Technology; and in the ranks such men as Allen Hazen and George W. Fuller, Members, Am. Soc. C. E.

While this is not an address on sewage purification, I cannot refrain from stating that the experimental work for the purification of sewage and water was planned and executed under the personal direction of Mr. Mills, who, although serving on an unpaid board, gave to this work the larger part of his time for several years.

The examinations of the water supplies of the State, with which the speaker was more intimately associated, and the continued analyses of the waters, proved the great value of the co-operation of engineers, chemists, biologists and bacteriologists, as was recognized by all who participated in the work. These advantages may not be as fully appreciated at the present day, when there is no difficulty in finding individuals who are well acquainted with the engineering, chemistry, biology and sanitary history of water supplies, and who are in many cases experts along more than one of these lines, but men with such knowledge did not exist twenty years ago, and that they are available to-day in this country is to be credited to a large extent to the work of the Massachusetts Board of Health.

Continuing with the subject of the co-operation of engineers with those in other branches of applied science, I wish to express my indebtedness to a skilled geologist, who, from extended examinations at the surface of the ground, with the assistance of only a limited number of borings, predicted with a high degree of accuracy the character of the rock to be encountered in different parts of several aqueduct tunnels; who, in extensive boring operations assisted by his advice as to the best method of obtaining trustworthy samples of the material penetrated, and who, by painstaking work and the resources of the geologist, interpreted the results of some important borings more accurately than they could have been interpreted by the engineers.

It is perhaps unnecessary to refer to the importance of consulting an architect when important buildings are to be erected, because it is now the general rule to do so, but there are more exceptions to the rule than there should be.

It is less common for engineers to consult a landscape architect, as many have faith in their ability to produce pleasing results with geometrical forms. The results of such co-operation in my own experience have been particularly satisfactory, as the

DAM NO. 2 AT FELLS RESERVOIR.

carrying out of a landscape architect's plan, made before beginning work, has resulted in pleasing effects at little additional cost, and I will refer at some length to the results of co-operation of this kind in connection with the construction of distributing reservoirs.

Plates XXX to XXXIII explain more clearly than words alone some of the effects produced by the landscape architect's design.

Plate XXX shows a highway along the northeasterly border of a distributing reservoir having an area of 300 acres, which was formed by deepening a natural pond and raising it 9 ft. above its original level. The highway before it was raised had much less pleasing lines than afterward, as it was laid out in nearly straight stretches, and on the water side the bank sloped directly downward from the guard fence at the side of the roadway. In the executed work, the roadway was laid out on easy curves, was kept further from the water, and uniformity of lines and material were avoided, with a view to suggesting natural rather than artificial effects.

Plate XXXI shows the landscape treatment where another highway was necessarily carried across a shallow arm of this reservoir on an embankment. As the portion of the reservoir on the left of the embankment would have been undesirable for water supply, even if deepened, it was left for park purposes, and entirely shut off from the reservoir, which is on the right of the embankment. The advantages of this landscape treatment over a formal treatment with an embankment having a uniform section, steep slopes and guard fences, will be readily appreciated.

Plates XXXII and XXXIII show two of the dams of an artificial reservoir, having an area of 8 acres, located near the summit of elevated land where Nature had formed in part a rock-walled basin requiring only five dams to fill the gaps between these walls. All parts of this reservoir were treated so as to give natural effects. Especial attention is called to Plate XXXIII where, both for landscape effect and to dispose of the large amount of surplus material excavated from the reservoir, a hill as high as any of the surrounding land was created. Buried in this hill is a concrete core-wall extending to a rock foundation, with a rolled embankment of selected earth on both sides of it. In looking at this reservoir it is somewhat difficult to realize that it is not a natural pond.

Both the reservoirs above mentioned are a part of the Metropolitan Water System of Massachusetts, and are located in a nearly natural woodland park.

The relative advantages of municipal and private ownership of public utilities is a live question in municipal circles at the present time, and in the discussion of this subject water supplies furnish the most valuable object lessons.

In the days of Greece and Rome, the aqueducts and other works were built by the kings and emperors, and operated under their direction. The question of municipal ownership arose in more modern times, when the municipalities became wealthy and to a large extent self-governing.

The Hampstead Heath Water Supply, for a portion of London, was built by the municipality and finished in 1590, but about one hundred years after, in 1692, it is recorded that the municipally owned water supplies taken from sources within 5 miles of London were sold to the Hampstead Water Company. After this sale London was supplied by private companies until 1904, when all the immense works for supplying water to that city were again taken by the municipality.

In America the water-works of Philadelphia were municipally owned from the beginning, but the earlier supplies for New York and Boston were introduced by private companies.

There has been for a long period of years a movement toward the municipal ownership of water supplies, and most of the larger water-works in America and in Europe are so owned. The municipally owned water-works, as a rule, have been successful in all respects.

In this address I have endeavored to give a brief account of the development of water supplies and water-supply engineering, covering both water supply as a whole and in some of its more important features.

I have found it necessary to call your attention to the great waste of water as one of the discreditable features of American water supplies, but I have indicated that the responsibility for this lies with the governing bodies of our cities and not with the

PLATE XXXIII.
TRANS. AM. SOC. CIV. ENGRS.
VOL. LVI, No. 1025.
STEARNS ON
WATER-SUPPLY ENGINEERING.

Dam No. 4 at Fells Reservoir.

water-works engineers. As a result of the great waste of water, however, the engineer has had occasion to build works of unprecedented magnitude, and he has built dams, aqueducts and other water-works structures so well that there have been few failures to record. He has installed pumping engines which are unexcelled in efficiency and capacity, and the works built, as a rule, have been financially successful under municipal as well as under private ownership.

The quality of the water supply of many American cities has been far below the proper standard, but improvements have been effected so rapidly within the past few years that it can be confidently claimed that under the leadership of the engineer and sanitarian a high standard in this respect will soon be reached.

Frederic Pike Stearns

Frederic P. Stearns, born November 11, 1851, died December 1, 1919, was an acknowledged leader in the field of water shed and water system engineering.

A native of Calais, Maine, he went to Boston in 1869 seeking employment after graduation from high school. His inclinations were already clear, and his first job was in the office of the Boston City Surveyor at the time when municipal engineering and large water systems were just beginning. From the start it was diligent application to his work and self-education that carried him to the top of his profession. His university degrees were honorary, for he never attended college; a Master of Arts from Harvard in 1905, and a Doctor of Laws from the University of Pennsylvania in 1906.

In this pioneering era of municipal engineering in Boston, there was a remarkable group of talented engineers that included J. P. Kirkwood, E. S. Cheseborough, J. B. Francis, Joseph P. Davis and Hiram F. Mills. It was with this inspiration that Mr. Stearns began his professional career; and his advance was rapid. At age 21 he was an assistant engineer for the Engineering Department of Boston on the preliminary investigations for the Sudbury water supply to the City of Boston. Two years later he was engineer in charge of the construction for a portion of that project. In 1880 he started construction work on the Boston sewerage system, a project noted for its adaptation to conditions. The tunnel under Dorchester Bay and the outfall works were constructed under Mr. Stearns' direction as Assistant Engineer under the City Engineer, Joseph P. Davis. Mr. Stearns was next engaged as chief engineer on design of the Stony Brook improvement, a pioneer project involving the scientific study of storm water problems.

In 1886 Mr. Stearns became Chief Engineer of the Massachusetts Board of Health. The State Board had just been reorganized and charged with the oversight and care of the inland waters of the state, with powers to advise municipalities on all matters relating to water supply, drainage, sewerage, and the p-evention of pollution of inland waters. Mr. Stearns, now age 35 years, contributed significantly to the outstanding reputation that the accomplishments of the Board gained for it. His studies of water supplies set a new standard for thoroughness and sound judgement; and his tact in dealing with officials, and respect for his opinions, secured public confidence in the Board. Particularly notable were his studies of stream flows and storage, establishing the principles of economical development of watershed yields.

As Chief Engineer for the State Board of Health, he conducted important investigations relative to a proposed sewerage system for the Mystic and Charles River Valleys. His plan was adopted. He served as consulting engineer for the construction of the Charles River Basin which converted a foul tidal estuary into an outstanding fresh water recreation facility. In this work he was associated with the famous landscape architectural firm of Olmsted, Olmsted & Eliot. This association may well have influenced Mr. Stearns' thinking, because his subsequent engineering creations were noted for their environmental beauty and the blending of engineering structures into their surroundings. Possibly his greatest achievement of this period was the plan for the waterworks system to supply Greater Boston. On acceptance of his plan, he resigned his position with the Board of Health to become Chief Engineer of the Metropolitan Water Board which was created to carry out the project. This great work was carried out between 1895 and 1907. For difficult problems solved, novel features and beauty, few water works projects have matched this one. This accomplishment particularly was the reason for the academic honors that Mr. Stearns received in recognition of his contributions to the advancement of engineering.

After his retirement from the Metropolitan Water Board in 1907, Mr. Stearn served as consulting engineer on projects within his special interest fields. These included the water supplies of Baltimore, Los Angeles, Winnipeg, Hartford, Rochester, Worcester, Providence and New York. He was also consulted on the Chicago drainage Canal, the Mississippi Power Dam, other projects in the West and Mexico, and on the Panama Canal.

His influence was felt by other engineers with whom he was associated, especially through his activities in the New England Water Works Association, and the American Society of Civil ENgineers of which he served as President in .

AMERICAN SOCIETY OF CIVIL ENGINEERS.

INSTITUTED 1852.

TRANSACTIONS.

Paper No. 1059.

WATER SUPPLY.

An Informal Discussion at the Annual Convention, July 10th, 1907.

SUBJECT FOR DISCUSSION:

"Is it better policy to purchase and control water-sheds, thereby preventing the pollution of impounding reservoir supplies, or to suffer a certain amount of pollution of such supplies, relying upon filtration to correct the effects thereof?"

BY MESSRS. GEORGE W. FULLER, F. HERBERT SNOW, D. D. CLARKE, GEORGE A. SOPER, WILLIAM J. BALDWIN, GARDNER S. WILLIAMS, J. T. NOBLE ANDERSON, WILLIAM B. FULLER, J. WALDO SMITH, PABLO SOLÍS AND OCTAVIO GUZMÁN, S. BENT RUSSELL, M. R. SHERRERD, F. W. BLACKFORD, GEORGE A. JOHNSON, AND GEORGE W. FULLER.

Mr. G. W. Fuller.

GEORGE W. FULLER, M. AM. SOC. C. E. (by letter).—Various local conditions require consideration, in deciding this question. This is particularly true as to the quality of water which it is desired to obtain, which quality must meet, or ought to meet, certain rational requirements of local or State health authorities.

It is undoubtedly true that water supplies which were believed to be good 10 years ago are not in all cases now regarded as satisfactory in quality. As to water supplies of good quality to-day, there are strong reasons for believing that some of them will not necessarily be thus regarded a few years hence. We are in a transitional stage as regards the quality of public water supplies. Advanced positions taken by various health authorities and by public opinion, without doubt, now make it urgent for the engineer to be keenly alive to sanitary requirements, in order that public water supply projects may keep abreast of the times.

Mr. G. W. Fuller.

The basic thoughts or premises should first be stated, in order to facilitate discussion, and, as they appear to the writer, they may be outlined briefly as follows:

1.—It is essential that the water from impounding reservoirs, or from any other source, shall be substantially free from disease-producing germs.

2.—A satisfactory public water supply should be of good appearance and free from objectionable quantities of mud, turbidity, and vegetable stain.

3.—A satisfactory water supply should be free from objectionable tastes and odors.

4.—Impounding reservoirs are of much assistance in the purification of water through natural agencies, as regards the removal of bacteria, turbidity, and peaty color. Their agencies are not infallible, and their efficiency varies much with different conditions as to rainfall, temperature, climate, storage period, and other physical factors.

5.—The elimination of pollution through the purchase of land is highly desirable, but in many places is prohibited by its cost. Sanitary control through inspection is helpful, but difficult of accomplishment under existing laws and customs.

6.—A well-designed and well-operated filter plant, with all its various appurtenances, such as efficient aeration, sedimentation, and coagulation, will allow a satisfactory water to be obtained from practically any impounded-water supply, although, within certain limits, the better the quality of the unfiltered water, the better and cheaper will be the filtered water.

Protection by Impounding Reservoirs from Disease Germs.—Where a reservoir has a capacity equal to the volume removed from it during a period of many weeks or months, it is obvious, from our present knowledge, that a large proportion of the disease germs will disappear from the flowing water. These germs have weight, and subside in fairly quiet water. According to laboratory experiences, disease germs live much longer in reservoir water than in sewage or highly polluted waters.

In all waters there is quite a rapid dying out of disease germs during the first week or so, and, with present laboratory methods, it is usually very difficult to detect the presence of any of these germs after they have been in water for two or three weeks. This does not prove, however, that all the germs have died out. On the contrary, the evidence seems to be quite strong that some of the more hardy ones, perhaps only a fraction of 1% of the total number, will survive for a long period, presumably months and years. Otherwise, it seems to be impossible to explain the continuance of water-borne diseases in certain districts.

While our knowledge of the passage of disease germs through large impounding reservoirs is by no means precise, our information is such that we cannot ignore certain physical conditions limiting the efficiency of the natural agencies there at work. Among such are the stirring up of the bottom layers of deep reservoirs, due to vertical circulation during each spring and autumn, and the effect of stratification and of various currents, which, under some conditions, no doubt bring the entering water to the outlet in an abnormally short period of time. Mr. G. W. Fuller.

It is futile to go into this subject in great detail here, but it is to be pointed out clearly that because a reservoir may hold 100 days' supply it by no means follows that the numbers of disease germs passing through the outlet are always as low as would result from the stay of these germs for 100 days in the reservoir water. Obviously, the protection from impounding reservoirs increases with the period of effective storage.

There are other factors also to be considered. Among them may be mentioned heavy rains, which sometimes wash into the impounding reservoir the accumulated filth of weeks and months, and also that such pollution may at intervals occur when the reservoirs are drawn down quite low, and when the opportunity for purification by natural agencies within the reservoirs is abnormally small.

Let us next turn to the question of how much pollution may suffice for impounding reservoir waters to produce disease among the consumers. The evidence is quite striking that the dejecta of one or two persons suffering from typhoid fever or other serious intestinal disorders may pollute surprisingly large volumes of water. This is particularly true when such dejecta accumulate during the winter months and are suddenly washed into the stream or reservoir when there is a general thaw. Experiences at Plymouth, Pa., New Haven, Conn., York, Pa., Scranton, Pa., and other places, show that the danger from a scattered population on the water-shed may at intervals be far greater than was considered possible a few years ago.

Notwithstanding its shortcomings, the impounding reservoir has been and continues to be a tower of strength in municipal sanitation in many places. This becomes quickly apparent upon comparing the typhoid death rate (about 20 per annum per 100 000 population) in New York, Newark, Jersey City, etc., which is as low as in numerous other cities more or less similarly situated, and having a public water supply either from underground sources or of filtered river water.

Filtration of upland water from sparsely populated areas, however, produces still lower death rates in some cases, as shown by the experiences of Paterson, N. J.

Mr. G. W. Fuller.

Mortality statistics as to typhoid fever and other water-borne diseases are frequently only a very rough measure of the quality of a public water supply. This is particularly true of waters which are of fairly good character. Pollution may exist at times without its presence being recognized. This is probably true at intervals in the case of the Croton supply. Early in the spring of 1907 there was a small outbreak of typhoid fever in New York City, stated by Health Commissioner Darlington to have been caused by Croton water.

One of the weaknesses of the impounding reservoir is that its inefficiency cannot always be detected or corrected in time to avert danger. This accentuates the importance of more attention being given to one or both of the procedures advanced in this theme for discussion, namely: the prevention of pollution on the water-shed, or its correction by filtration.

Protection by Impounding Reservoirs from Unsatisfactory Physical Condition of Water.—Sedimentation in large impounding reservoirs usually produces sufficient clarification, and it is only at times of exceptionally heavy rainfall that the water is muddy when delivered to the consumers. As compared with the waters in large rivers, especially in the South and West, the impounded supply is quite satisfactory in this regard.

The remaining turbidity in the water drawn from an impounding reservoir, however, is usually due to very fine particles, dark in color, which produce what is called a "dirty appearance" in the water; this is frequently complained of much more than would be expected by the users of muddy river waters.

In the case of waters drawn from drainage areas containing more or less swampy tracts, the impounding of the water does much to bleach it by the aid of sunlight, and to lessen materially the brownish tint due to organic matter extracted from leaves and other vegetation.

It is with regard to tastes and odors that the most disagreeable physical conditions are likely to be encountered in the water of an impounding reservoir. These may be due to the decomposition of organic matter on the bottom of deep reservoirs, or to the growth and disintegration of various small forms of vegetable and plant life which make their appearance in reservoir waters in a more or less irregular fashion and in a manner not now fully understood.

The stripping of the bottom and sides of reservoirs does much in the early life of a reservoir to reduce the likelihood of bad tastes and odors due to vegetable growths. It is not an absolute safeguard against these troubles, however, and it is to be further pointed out that the effect of stripping is not wholly permanent in its nature. In fact, present evidence indicates that soil washings, in course of time, perhaps 10 or 20 years, will largely obliterate the effect of stripping. Mr. G. W. Fuller.

When it is considered that a public water supply ought to be thoroughly "clean," it can hardly be maintained, in the light of modern requirements, that an ordinary impounding reservoir can give uniformly a thoroughly satisfactory supply in the absence of filtration, or of aeration, or of both.

It is true that swamps may be drained and various other expedients resorted to, but it is not ordinarily possible, in the opinion of the writer, fully to meet modern requirements as to the physical conditions of water through the agencies which act in the ordinary impounding reservoir. There are, of course, exceptions to this rule, and some exceptions will no doubt continue for a long time to come. The experiences and conclusions, however, at several places where bad tastes and odors prevail, show that, independent of pollution upon a water-shed, filtration is highly desirable and well worth its cost. The fact that trouble from microscopic growths has not yet been experienced in any given impounded supply is no proof that difficulties will not arise in the future.

Protection by Filtration.—For the most part, filtration has been practiced in America in connection with muddy and polluted river waters. For impounded water supplies, sand filters have been used successfully for two of the gravity supplies at Reading, Pa., the Ludlow supply at Springfield, Mass., and have been introduced recently at Bar Harbor, Me. Impounded water supplies have been treated satisfactorily by mechanical filtration at a number of places, among which may be mentioned Lexington, Ky., Middletown, N. Y., Norfolk, Va., and Charleston, S. C.

Available evidence shows clearly that by filtration a water can be made thoroughly satisfactory as to freedom from disease germs and objectionable tastes and odors, as well as of satisfactory appearance. Modern filter plants, both in their construction and their operation, require close adjustment to existing local conditions, and are bound to vary somewhat in their detailed arrangements. Filtration, in this discussion, is viewed in its broadest light, and with it is included efficient aeration and the use of coagulants when necessary.

As to the cost of filtration, that of an installation, complete, with all appurtenances, usually ranges from $15 000 to $30 000 per million gallons daily capacity. Usually, the figures for impounded reservoir waters would be nearer the lower figure, although there are special conditions in some places which require the higher figure to be exceeded. The cost of operating filters of moderate size usually ranges from $1.50 to $3.50 per million gallons for reservoir water, exclusive of pumping. Combining the operating expenses and capi-

Mr. G. W. Fuller.

tal charges, taken at 5% per annum, there is obtained, for fairly representative conditions, the sum of $7 per million gallons. On the assumption that 100 gal. of water are used daily by each inhabitant, this would mean a total annual expense of about 25 cents for each inhabitant.

Filters ordinarily remove from 97 to 99.5% of the bacteria in the applied water. Where bacteria have been removed to a large extent by sedimentation in impounding reservoirs, the numbers are sometimes so low in the unfiltered water that the above percentages are not obtained. The residual number of bacteria, however, would be very low, and the water would be fully equivalent in hygienic quality to the above stated percentages of removal as we understand them in connection with practical operations in the filtration of river waters.

Filters are not germ-proof, and none of their warmest advocates claims them to be. Their efficiency has proved a tower of strength in municipal sanitation in the past—even greater than that of the impounding reservoir—notwithstanding the fact that a small fraction of the applied bacteria passed through the filters and appeared in the water supplied to the consumers.

Obviously, the more polluted the unfiltered water is, the less safe will be the filtered water. The logical question in connection with this discussion is this: To what extent can water-sheds be purchased and controlled, with the result that the water is made of thoroughly safe quality, and under conditions where the investment ~~on~~ the water-shed is preferable to an investment at a filter plant?

Two fundamental things are to be borne in mind: First, that, all things being equal, a pure water is better than a purified water; and, second, that by adjusting the rates of filtration, by double filtration, and by the use of germicides, it is perfectly practicable for filtration experts to deliver to the consumer almost any grade of filtered water that is wished, provided the purification works are built and operated strictly in accordance with the best available knowledge of sanitary science.

This brings us to the question of relative safety, and the reliability of human efforts, both as regards the control of water-sheds and faithfulness and efficiency in the operation of filtration works.

Comparative Cost and Reliability of Preventing Pollution on the Water-shed and of Removing Pollution by Filtration.—Data are available to permit, for any specific case, a fairly close estimate of the comparative relations of these factors. It is difficult, however, to generalize on this question, owing to the varying local conditions, and they will be treated in this discussion in an illustrative way.

Mr. G. W. Fuller.

The total cost of efficient filtration may be taken at an average figure of about $7 per million gallons per day, including interest, sinking fund, repairs, replacements, and operating expenses. This is equal to about $2 500 per annum, corresponding to about $50 000 as the sum required for the maintenance of filtration works of a capacity of 1 000 000 gal. daily in perpetuity.

A water-shed area corresponding to a dry-weather yield of 1 000 000 gal., of course, depends much upon the rainfall, the evaporation, the relative capacity and water surface of the impounding reservoirs, etc. For purposes of illustration, let us assume that 1 000 acres of water-shed area are required for the development of 1 000 000 gal. of water daily. This means that, for the purchase and control of water-sheds, there would be available about $50 per acre, in order to have the cost of such treatment approximate that of filtration.

Where land is cheap, the water-shed area small, and the reservoir capacity large, there is merit in purchasing the entire land in some cases. For projects of moderate size, the writer believes that the likelihood of such purchase doing away permanently with the needs for filtration is too remote to be considered seriously.

Based on the above assumptions as to cost, it is seen that, for large water-sheds, where there are numerous villages, fertile farming lands, and desirable locations for country residences, it is hardly feasible to purchase the entire water-shed for anything like the sum which would permit efficient filtration to be maintained for all time.

If the costs were equal, and only one procedure were to be adopted, it is the writer's opinion that filtration would be preferable, not only on hygienic grounds, but for the sake of assuring a water supply of suitable physical character at all times.

In this it is to be borne in mind that filters have been installed recently at Bar Harbor, Me., notwithstanding the fact that there are no regular residents upon the water-shed, although it is stated that there are some three or four summer camps. Bad tastes and odors required the adoption of filtration.

As to the purchase of those lands upon which there are sources of objectionable pollution, and the control of the remainder by sanitary inspection, the writer is aware that this can be done with much benefit upon small water-sheds, but that it is frequently difficult to secure satisfactory results on large water-sheds by purchase and control.

The City of New York has undoubtedly benefited to a considerable degree by the elimination of objectionable sources of pollution upon the Croton water-shed, through the purchase of large numbers of small tracts of land, especially along the watercourses. However, the writer is keenly appreciative of the limitations to

Mr. G. W. Fuller.

these methods as applied to the Croton water-shed of 360 sq. miles of drainage area.

Italian settlements on low, wet land may be purchased to-day, but they will certainly spring up again to-morrow in almost equally objectionable locations.

Sanitary inspectors may provide thoroughly sanitary arrangements for the removal of fecal discharges, but of what avail are they with a population whose habits are such that they will not use these sanitary conveniences, and whose lack of education is such that they persistently and regularly resist efforts to protect both themselves and the users of the water from the drainage area upon which they are resident?

Several things must be corrected before the purchase and control of water-sheds for large projects can approach the efficiency of filtration when conducted with moderate intelligence and skill as practiced in numerous places to-day.

A campaign of education is required, not only for the Italian on the Croton water-shed, but for many others whose views would no doubt respond to educational efforts to correct their ignorance and indifference. Such corrections will assume practical shape largely in proportion to the vigor with which wise sanitary laws are carried out promptly, and with a firm hand, to administer proper punishment in an equitable manner, and about which there can be no dispute as to the division of responsibility between State departments of health, municipal departments of health, and municipal water departments.

The history of sanitary control of water-sheds in America, in the opinion of the writer, shows that it does not offer nearly the protection to the water consumer that is afforded by modern filtration. Sanitary control has been of great help in numerous places, but there are other instances where laws are inefficient, residents on the water-shed careless and indifferent, and where efficient results are more or less confined to the period when the inspector is actually on the premises. Better laws, better and larger organizations for sanitary inspection, and a better understanding by the residents of what is required can no doubt bring about improved results in the future.

While the writer is not in favor of sanitary control to the exclusion of filtration, he is clearly of the opinion that more stringent sanitary laws than now exist in most places should provide for the removal of all gross pollution, and that simple, practical methods should be gradually introduced to eliminate uniformly the lesser pollutions in the order of their importance.

Mention should be made of the reliability of filtration as actually carried out in practice. A decade ago many filter plants were such

Mr. G. W. Fuller.

in name only. The last few years have seen a marked change for the better. Practically all the larger filters are now well designed, and are arranged so that they can be controlled readily by intelligent, faithful attendants. Most important of all is the almost universal custom of installing laboratories at the filters, and of placing the operations under constant technical supervision. In no branch of sanitary engineering has more progress been made lately than in the efficient management of water filters for the larger cities.

In the future, such management will certainly continue, as it will doubtless be obligatory to furnish frequent and full records of the results of operation to supervising State authorities.

Compared with sanitary inspection, scattered at different points over a water-shed, and in the hands of various individuals, it appears to the writer that there is far less likelihood of inefficiency in the operation of filters, centrally located, with the management under a single roof, and in the hands of a single person.

European Tendencies.—In Germany, where for a dozen years filtration has been practically compulsory for surface-water supplies, there are only a very few important cities which derive their water from impounding reservoirs. In Great Britain, however, there are many such supplies, and it is significant that the majority of the engineers having charge of new impounded-water supplies have provided not only for filtration, but also for the purchase of the water-shed, where this can be done at moderate cost. Fortunately, such purchase has been practicable, and has been carried out, at the new water-works of Liverpool, Birmingham, Swansea, Cardiff, Manchester, Edinburgh, etc. Land on these drainage areas is comparatively inexpensive, and the writer thinks that there is no room for doubt about the wisdom of eliminating at nominal cost all sources of pollution from a resident population.

There are various reasons why it is usually difficult in America both to filter and to purchase or control the water-shed, notwithstanding the desirability of doing both, from a theoretical standpoint. It is even necessary in some cases to install filters for polluted streams, when impounded-water supplies could be availed of as far as physical conditions are concerned. Under such circumstances, it is the writer's practice to arrange the filters so as to serve for the upland supply when financial and other conditions will permit its adoption.

Conclusions.—1.—The purchase and control of water-sheds is a less efficient procedure than filtration, as regards both the sanitary character of the water and its physical condition as to appearance, tastes, and odors.

2.—For projects of much magnitude, filtration is generally the cheaper.

Mr. G. W. Fuller. 3.—Sanitary control of a water-shed is of much assistance, but it is usually a difficult matter to manage well at all times for large water-sheds, under existing conditions as to sanitary laws, and as to the views and responsibility of rural population. Better laws are generally needed.

4.—The purification of water due to natural agencies in a large impounding reservoir makes its filtration easier, cheaper, and more effective than in the case of the ordinary river water. Purification plants can be installed to insure almost any degree of efficient purification that is desired.

5.—Notwithstanding the efficiency of filtration at reasonable cost, there is a growing tendency in America to provide as good sanitary conditions on a water-shed as practicable. As a rule, it is not believed that the purchase of the entire water-shed is advisable, on the ground that the money involved can be expended better in other ways.

Mr. Snow. F. Herbert Snow, M. Am. Soc. C. E. (by letter).—The sanitary conservation of public water supplies is a subject which forced itself upon the attention of the General Assembly of the State of Pennsylvania in 1905, resulting in the enactment of a law entitled "An act to preserve the purity of the waters of the State, for the protection of the public health," under which the Commissioner of Health is administering that branch of State medicine relative to public water-works and sewerage in a way to prevent ultimately a recurrence of epidemics of a water-borne character.

During the winter of 1906-07 three notable epidemics occurred in Pennsylvania, namely, at Scranton, Warren and Kittanning. Prior to this time the State had furnished the classic typhoid fever epidemic at Plymouth, and later at Butler. The facts in connection with these cases and the lessons they have taught should be clearly in the mind of the expert called upon to advise with respect to the amount of sewage pollution permissible on a water-shed.

The Scranton typhoid scourge came from an unfiltered surface supply of water derived from a mountain water-shed on which were located three villages and two railroads. The disease struck the town suddenly, lasted about 6 weeks, totaled more than 1 000 cases and rolled up a record of more than 100 deaths. No infection could be found on the water-shed. The poisoning of the water must have been accidental, but, nevertheless, it was certain. Typhoid bacilli were discovered in one of the reservoirs.

The lower basin, from which the gravity supply mains extend to the town, has a capacity of about 100 000 000 gal. The main impounding reservoir is 5 miles above, and has a capacity of more than 1 000 000 000 gal. A short distance above is a third impounding reservoir having a capacity of about 400 000 000 gal. Extending for several miles along and across the main stream, and by the shores of the three reservoirs, are the two railroads. Droppings from the passenger coach closets might easily have been washed from the roadbed into the city supply. Mr. Snow.

Although these and numerous other permanent menaces had existed on the water-shed for years, Scranton had never before been afflicted by a water-borne disease. Whatever pathogenic pollution had reached the reservoirs was either effectually destroyed by natural agencies or rendered impotent. Retention in the impounding basins must have been one of the formidable barriers against the entry of infection into the water pipe system. No doubt is entertained about such infection having been present in the city supplies prior to the epidemic. The topographical evidences are conclusive.

It was not due to improper operation of the water-works plant, or to any apparatus or appurtenance thereof, that the flood gates of disease were let down in this instance. Simply this: there came a day when all the natural barriers proved of no avail. Without any change in the physical conditions on the water-shed, and without warning, Nature's safeguards went down. What had proven for thirty years or more to have been an adequate plant for furnishing a pure and wholesome supply suddenly became inadequate, thereby transmitting disease and death to the community. The origin of the infection will probably never be known.

An intelligent mind cannot reflect on the history of this catastrophe without being impressed by the lesson it teaches. Barriers against water-borne diseases, to be effective under all conditions, should be capable of manipulation and control. The water company had purchased several hundred acres along the streams and around the reservoirs, and had attempted to control the water-shed by a system of patrols constantly maintained. Many nuisances were thus abated and some prosecutions made to prevent the pollution of the impounding reservoir supplies. During the epidemic, the State assumed control of the water-shed, and put in force rigid rules and regulations for the sanitary protection of the waters thereof. There are scores of public water-works systems to-day where the sources of supply are as likely to be infected, accidentally or otherwise, as the Scranton source, and even more so. Many human lives are bound to be sacrificed before adequate and universal preventive measures shall have been adopted.

At Warren, beginning on December 8th, 1906, an epidemic of gastro-enteritis raged for four days, during which 1 200 patients called for medical attention, and about 600 more were reported bedridden, but did not receive professional care.

This borough takes its water supply from driven wells, in the

Mr. Snow. town, on the banks of the Allegheny River, the waters of which are polluted by sewage. Investigations by the Department of Health pointed conclusively to the water as the medium of transmission of the poison. From the nature of the ailment, it was premised that the infection must have been in concentrated form.

The cause was discovered to have been the sudden rising of the river and the back-flooding of sewage through a drain pipe and ultimately through loosened joints of the piping of the driven-well system and into the town's supply. Attention to the sewer, which was in a leaky condition, and to the joints and packing of the suction pipes, which joints and packing had been loosened by alterations in the main suction pipe, would have prevented trouble. Too much care cannot be exercised with respect to the condition of the apparatus by which ground-water is obtained for a public supply, when such apparatus is located on the banks of a sewage-polluted river in which freshet heights inundate the surface of the land upon which the wells are located. The danger exists if there be sewage in the river, or a conduit, or a receptacle, or anywhere whence it may be transmitted to the drinking water. The lesson taught by this experience is that a ground-water supply, regularly proven by tests to be of superior quality, may suddenly become contaminated from a known remote source, and be prejudicial to the public health. The remedy, therefore, if it be practicable to do so, is to remove the remote danger, in other words, to pay attention to the proper disposal of sewage.

The sewers of Warren discharge into the Allegheny River about 90 miles above the borough of Kittanning, in which the source of public supply is the same river. The water was filtered, and was supposed to be purified. However, expert investigation revealed the fact that the plant had not been operated intelligently, nor was it equipped to render satisfactory treatment of the water possible. Yet the public was led to believe that the system of purification as installed and operated was a sufficient safeguard.

Several days after the Warren outbreak a similar epidemic of gastro-enteritis broke out in Kittanning. It lasted more than 8 weeks and numbered in that time upwards of 4 000 cases. The physicians of the borough reported that the vast majority of citizens were afflicted and received medical attention.

Pathogenic pollution from many municipalities on the banks of the river above Kittanning may arrive at the water-works intake at the latter place in condition capable of causing disease. The Warren infection might have been transmitted to Kittanning and have produced the epidemic there. The coincidence of outbreak in the two places attracts attention, although a local source of infection is more probable. But, besides the bowel trouble, there were more than 100 cases of typhoid fever in Kittanning. This infection is believed to have been introduced through the medium of the river. Mr. Snow.

This case demonstrates how utterly inefficient may be the effort of those who, misinformed as to the essentials of water filtration, try to safeguard human life. Here was a water purification plant, installed for the purpose of preventing a recurrence of typhoid fever, totally inadequate to perform the service expected of it by the public, but, nevertheless, serving to foster a false sense of security.

In such a case, where the source is always dangerous, the interests of the public health demand competent design and skilled supervision of the construction of the purification works at first, and thereafter, capable management of its operation, and efficient maintenance.

The business of furnishing drinking water to the public is bound to be regulated by laws, passed to be enforced by central administrative power. The courts are beginning to give a different meaning to the words, so frequently appearing in water-works franchises, "pure and wholesome supply." Practical methods have been evolved whereby it is possible now to render water purer than it may be possible to obtain it in a raw state. The consumer is entitled to the kind of article he purchases, that is, a "pure and wholesome supply," and a Pennsylvania court has recently held that primitive methods of water purification, whereby a slightly better water than exists in the stream is furnished to the consumer, is not sufficient as long as that water is prejudicial to the public health.

The Butler epidemic, of the winter of 1903-04, affords a different lesson. At that place the public water supply was adequately filtered by a plant properly designed and operated. The source was from a rolling farming district, yielding turbid waters at times, and subject to possible sewage pollution. The filters had been installed to eliminate all danger and produce a clear water. This they did. The plant was one of the advertised assets of the borough; therefore, when the town woke up one day to the presence of considerable typhoid fever in the community, no one, not excepting even the newspapers (so alert usually to scrutinize critically a public utility plant), seriously suspected the water supply; and, remarkable to relate, not until 1 000 people were stricken and in bed was State aid called for and the true origin of the scourge discovered. The disease did not come in a day, but was cumulative.

The filter plant had undergone repairs just prior to the inception of the epidemic. During this period of repairs—several days—crude creek water was introduced into the street mains. There were several cases of typhoid fever on the water-shed, in a dwelling on the banks of the stream. The dejecta from the patients were

Mr. Snow. thrown out on the surface of the ground, where nothing prevented the poison from being washed into the creek, and hence within a few hours it arrived in the homes of the unsuspecting, confiding and wholly helpless water consumers.

Had the local or State health authorities been made aware of the first outbreak of the disease, or of the original cases on the water-shed, the history of the Butler epidemic might never have been written. But, as it was, 1 348 cases were recorded and 111 deaths.

This catastrophe produced one good result, namely, the establishment of the State Bureau of Vital Statistics. At present, physicians are required to send in morbidity reports. A repetition of the Butler case is not possible. That is, a community cannot be gradually attacked by typhoid. The onslaught must be sudden and sufficiently intense to cause the harm at once, because the officers of the local and State health departments are ever on the lookout, and report to the central office the first appearance of any infectious disease.

However, systems of reports can only apprise of danger. The poison may be disseminated through a water-works system in a few hours, and the first infection may take place before knowledge of it is possible. To the initial infection at Scranton should be attributed about all the cases there. Had the disease escaped detection in its early inception, the epidemic would undoubtedly have far exceeded the proportions it assumed, and thus might easily have gone down in history as the most appalling typhoid scourge on record. Thus the early discovery and the shutting off of the infected water in this one instance alone probably saved to the community, besides human life, a money consideration greater than the amount expended annually on records by the State and city Departments of Health.

In another instance, Johnstown, by the prompt report of typhoid on the water-shed, was saved from an epidemic under circumstances otherwise assuring a sweeping disaster. The poison was intercepted and killed, and not a case occurred in the town. Unquestionably, the morbidity reports instituted by the Commissioner of Health will repay to the public the cost of maintenance of the Bureau, and this work has been made necessary in following out the policy of preventing pollution of impounding reservoir and other surface supplies, as far as it is practicable to do so.

The Butler lesson teaches that an efficient filter plant is not an absolute safeguard; that as long as there is poison in the source, the danger of an explosion of disease of a water-borne character exists.

The Plymouth epidemic of 1885 attracted the attention of the world's sanitarians. The case was unique at that time. A town,

Mr. Snow.

nestled at the foot of a mountain, furnished with a pure mountain water coming from an uninhabited water-shed—excepting one dwelling—was suddenly the scene of the terrible scourge of typhoid which numbered 1 104 cases and 114 deaths. The discovery of the origin was, for that day, a masterful work of detection, clear-cut and uncontroverted.

The owner of the single dwelling-house on the water-shed went to Philadelphia, contracted typhoid, came home, and was attended by a physician and nurse who did not know that the little stream at the foot of the bluff, on the edge of which stood the cottage, was the source of Plymouth's drinking water, and so they threw the poisonous stools out on the ground, where the dejecta remained frozen during the cold winter months; but in April, when the thaw and warm rains came, and the ice and snow all disappeared, the typhoid poison, along with everything else, was carried into the creek, and thence some of it found its way through the water pipes and into the town.

This was not the only supply to Plymouth. The Susquehanna River, receiving the sewage from Wilkes-Barre, 3 miles above, was used occasionally for parts of the town. This source was thought to be dangerous because it contained sewage, and suspicion had been directed to it for a long time. The public persisted in attributing the epidemic to the Susquehanna River water, until the proof of the true origin became conclusive. Although Wilkes-Barre's sewage constituted an immediate and always present menace, yet the citizens of Plymouth had not, prior to 1885, suffered from typhoid more than had other communities in the vicinity having sources of water supply which were not from the river. Contrary to the expectations of those interested in this case, came the proof that a constantly sewage-polluted river supply used by people who know the danger in the water, may not be as prejudicial to public health as a mountain supply free from pollution and having but one apparently insignificant menace upon the water-shed.

The disease held on through the summer months. Privy vaults abounded in the borough, also wells in close proximity thereto. People, mostly miners of foreign birth, would not heed the admonition to boil all water, and to disinfect all dejecta. Consequently, the wells were poisoned and the percentage of secondary infection was large.

The general lesson taught was startling and broad. Since one case of the disease imported could pollute the water supply of an entire town sufficiently to produce wholesale sickness and death, the officers of public and private corporations all over the country were startled at the proposition that no municipality taking drinking water from surface sources is safe.

Mr. Snow. The conclusion reached at that time was that the individual in the village or rural district must dispose of his excrement in such a manner as not to pollute, directly or indirectly, any surface or well water.

The remedy was accepted as ideal, and impractical of accomplishment. To-day, with the aid of advanced methods of water and sewage purification, of public hygiene and municipal sanitation, a very much nearer approach to the ideal is practicable than in 1885. To this end the Pennsylvania State Department of Health was reorganized by the law of 1905, and the responsibility for progress in this branch of State medicine was placed in the hands of the Commissioner of Health.

The instances herein cited serve as a foundation for the enunciation of the principle that no unfiltered surface water, in any way likely to be polluted by sewage, is absolutely safe for drinking purposes. The risk attendant on the use of it may be slight. Where the risk is great, the life and health assurance afforded by filtration of the water is demanded as a public safeguard.

Still further, filtration is not an absolute safeguard, more especially when the source is subject to sewage pollution. The interests of the public health demand that sewage pollution of public supplies shall cease. Standards of excellence in the design, construction, and operation of water filter plants are first essentials to guard the public health, but this safeguard should not put a premium on the use of the watercourse as a sewer by some up-stream municipality. On the contrary, the original pollution of the water must be minimized.

A year ago the proper authorities of the three sovereign States—New York, New Jersey and Pennsylvania—met to formulate a policy of co-operation with respect to the sanitary conservation of inter-state streams and their tributaries. It was unanimously agreed that water filters were not alone sufficient, but that both water and sewage purification plants are demanded, in the interests of public health.

The right of eminent domain in Pennsylvania has been taken away from public water companies. It has been decided to be contrary to public policy to permit such companies to acquire thousands of acres of land for protective purposes. Further, it has been observed that purchase and control of large water-sheds does not necessarily prevent pollution.

A certain quantity of sewage is bound to find its way into the waters of most inhabited areas, but the legalizing of a certain amount of pollution of such waters, relying upon filtration to correct the effects thereof, is a dangerous expedient to practice generally. Start with the proposition that, after all is done that may be done to keep sewage out of drinking water, the streams draining populous areas will still be unfit to use unless the water be filtered, and, using this as an argument against sewage purification works, grant licenses to towns to discharge sewage into the watercourses, and inevitably the pollutions must exceed the "certain amount." Mr. Snow.

The popular notion that it is permissible to discharge sewage into a stream which is not thereby made offensive to the sense of sight and smell is far from being dispelled. The local authorities of many towns are honestly dense with respect to this subject. It took an epidemic to teach the 1905 Board of Health of Nanticoke that the excrement of one individual could be transmitted many miles and kill those who drank the poison, although nothing might be visible in the water to indicate the danger. Yet Nanticoke is nearly opposite Plymouth. After the lapse of twenty years, the Plymouth lesson had to be repeated in the same district. It is not from the standpoint of nuisance, but from the standpoint of pathogenic pollution, that the campaign against putting sewage directly or indirectly into the waters of the State is being waged in Pennsylvania.

When the enormity of the crime of defiling water supplies appeals generally to everybody, there will be voluntary reforms. Meantime, the administration of law is a very pertinent matter. There is no public work of more importance to health than water purification and sewage disposal. Pennsylvania provides for central supervision of these utilities. No water-works or sewerage system can be legally extended or installed unless plans therefor have been submitted to and approved by the Commissioner of Health. Respecting the discharge of sewage, by a municipality, into any stream or body of water, the unanimous agreement of the Governor, Attorney-General, and Commissioner of Health is required. A penalty of $50 is provided for every day that a municipality pollutes illegally any of the waters of the State.

Dr. Charles B. Penrose, of Philadelphia, conceived the new State health laws, prepared the bills, and labored assiduously for their enactment. They are a worthy monument to a worthy man. The Honorable Samuel G. Dixon, who, as first Commissioner of Health, has had the task of organizing the new Department to put his conceptions of State medicine into operation, has inaugurated a policy which is bound to reduce water-borne diseases to a very large extent.

Every minor source of pollution on the water-sheds is to receive attention, and all menaces are to be permanently abated. Patrols by private and municipal corporations are to be required. The owner of each estate must cease to be careless about the disposal of sewage. Such disposal must be accomplished in a way unharm-

Mr. Snow. ful to others. Ordinary farm operations cannot be interfered with, but the effort is to confine pathogenic material within bounds. A health officer has been assigned to every municipality, to investigate, report, and execute orders.

Major pollutions, such as municipal sewerage systems, require more consideration. Often, the problem is one of finance. Slowly but surely the sewage of towns will cease to be discharged, untreated, into the waters of the State. First-class operation and maintenance of plants which have been constructed under State approval will be required. The necessity for State supervision is recognized, in fact, demanded. Without it, the "certain amount" cf pollution of surface water supplies would be a large and dangerous amount, and the moneys, therefore, expended in the construction of sewage disposal works would be lost to a considerable extent.

Mr. Clarke. D. D. Clarke, M. Am. Soc. C. E. (by letter).—That the purity of the water supply of any city is a matter of vital importance is so generally recognized that it seems almost superfluous even to state the fact; but, when it comes to the question as to the best method of securing and maintaining the purity of the supply, there is a diversity of opinion. The writer holds that it is better to secure such a supply as shall be of unquestioned purity, rather than attempt the purification of waters known to be contaminated in the slightest degree; but it is recognized that the financial question involved is often a controlling factor, making necessary the adoption of methods which otherwise would not be approved.

As an illustration of what it is possible to accomplish, at least in the more recently settled portions of the continent, the writer would cite the case of Portland, Oregon, a city of 150 000 inhabitants, which secures an abundant supply of pure water from an absolutely uninhabited water-shed situated on the western slope of the Cascade Mountains from 30 to 50 miles distant from the city.

The advantage of this supply, as a factor in the life and growth of the city, can hardly be over-estimated, and it is safe to say that it is with a growing appreciation of its value that the residents of Portland regard everything that pertains to it or affects it in any way; and hence they are fast coming to feel in a measure justified in felicitating themselves upon their good fortune in being permitted to dwell in a city possessing such an abundant supply of pure water.

It is now nearly twenty-five years since the question of municipal ownership began to be discussed, and surveys were made looking toward securing a water supply for the city from a mountain stream.

In 1886 a citizen's committee, numbering fifteen of the substantial business men, was empowered by the State Legislature to purchase or construct and operate water-works for the supply of the city. In the following year the Water Committee purchased the plant of the old Portland Water Company, which had been supplying the city with water from the Willamette River by pumps installed at a point on the river bank about 6 miles above the city; these works they continued to operate until the Bull Run gravity system was completed in 1895. Mr. Clarke.

The surveys for the gravity system, made under the direction of the late Isaac W. Smith, M. Am. Soc. C. E., who also constructed the works, developed the fact that the most available and altogether satisfactory source of supply was to be found in the water-shed of Bull Run River, in the foothills of the Cascade Mountains, eastward from Portland. The point finally selected for the intake is situated on the Bull Run River, about 5 miles from its junction with the Sandy, a tributary of the Columbia, and 30 miles from Portland, the elevation of the intake being 720 ft. above low water in the Willamette River. This point is in the edge of the foothills but well above the general level of the valley, while thence eastward the ground rises rapidly to the summit of the Cascade Range.

The entire western slope of the range is covered with a dense forest, and the work of exploration is difficult at all times. Little was known regarding the water-shed of the Bull Run prior to the time of making the surveys for the city. From the maps on file in the office of the United States Surveyor-General, the approximate area of the water-shed was known to be about 140 sq. miles. The existence of a lake near the summit of the range was also known, and but little other information was available concerning the region east of the point selected for the intake to the conduit. It should be stated, however, that prior to the selection of this stream, as the source of supply for the city, the main stream and the lake were explored by engineers in the employ of the Committee, and it was thereby determined that the flow of the stream was ample to provide for the future needs of the city.

In 1891 additional explorations were made by Mr. Frank T. Dodge, Superintendent of the Water Department, and, as a result of his recommendations, the Water Committee made formal application to the Department of the Interior for the withdrawal of the entire region as a Government reserve. This withdrawal was accomplished on June 17th, 1892; under the name of the Bull Run Forest Reserve, covering an area of 222 sq. miles and entirely surrounding the water-shed of the main Bull Run River.

Regarding this reserve the following statement is made:

"The United States will not sell any land in this reserve, nor permit timber to be removed from it, nor allow sheep or cattle to

Mr. Clarke. be pastured thereon. So long as the laws governing these matters remain unchanged, the water supply of Portland cannot be injured, providing the law to prevent forest fires on the public domain is properly enforced. The forest shades the snow which every winter covers the land deeply, and retards its melting; the undergrowth protects the soil and prevents it from being washed into the streams. The result is that no settling reservoir is needed, the water is always clear as crystal and free from sediment even during the greatest freshets. It is so cool that it does not need ice to make it palatable."

At the time the Bull Run reserve was established there were within its borders but few tracts claimed by private parties, and these were mostly held as timber claims by non-resident owners. Only two or three of the claimants ever made any attempt to build houses or make any permanent improvements in the way of clearing and cultivating the land. In a very few years the Water Committee succeeded in acquiring control of all improved claims, and has since continued the work of extinguishing private title to all lands within the reserve. At the present time the entire reserve above the intake is absolutely uninhabited.

From the beginning it has been realized that the greatest danger which threatens the water supply is that due to forest fires, which are very destructive during the summer and early autumn of especially dry seasons, these fires being usually due to the carelessness of hunters, fishermen and campers within the reserve or along its borders. In order that the danger from this source may be reduced to a minimum, the reserve is patrolled by forest rangers employed by the Government, and, in addition, extra patrolmen are frequently employed and paid by the Water Department, in order that a thorough inspection and isolation of the district may be maintained. This work is in charge of the officials of the United States Forestry service. In furtherance of this protection work a special Act of Congress, passed April 28th, 1904, prohibits all unauthorized persons from entering the reserve, all trespassers being liable to a fine not exceeding $500.

The low-water flow of Bull Run River is estimated to be from 65 000 000 to 70 000 000 gal. per day. If a larger quantity should ever be required, it would be a question of storage in Bull Run Lake or elsewhere. Rainfall observations, at the head-works, at the western border of the reserve, show a yearly precipitation of about 70 in., while, at the higher or eastern portions of the reserve, it is undoubtedly much more. It will thus be seen that the average run-off from the rocky and precipitous Bull Run water-shed, of 140 sq. miles, must largely exceed any probable requirements. It is estimated that Bull Run Lake alone will afford ample storage for many years to come. By a comparatively short tunnel, the lake could be drawn upon so as to supply an extra 100 000 000 gal. per day for the possible 60-day period when the stream is at its lowest stage. Mr. Clarke.

By special Act of the State Legislature, passed in 1906, the entire flow of Bull Run River is secured for the exclusive use of the City of Portland.

The following is an analysis of a sample of Bull Run water taken from the stream after a heavy rain:

Total Fixed Ingredients.

"2.7 parts in 100 000 parts by weight, or 1.89 grains per imperial gallon (of 10 pounds avoirdupois) consisting of:

	Parts in 100 000.	Grains per gallon.
"Silica	0.56	0.392
Oxides of iron and aluminum.	0.08	0.056
Calcium carbonate	0.39	0.273
Magnesium carbonate	0.27	0.189
Chlorides, sulphates and carbonates of alkali	0.40	0.280
Organic matter	1.00	0.700
Total	2.70	1.890

"The carbonates of lime and magnesia are in the form of soluble bicarbonates. The organic matter is mostly in the form of suspended vegetable substances and partly insoluble products of their decomposition. Microscopic examination of the sediment, and chemical tests of the water show the absence of deleterious organic matter. Only traces of ammonia were found, while nitrates and nitrites could not be detected, showing absence of nitrogenous organic matter. To sum up, the water ranks among the best on record for domestic use."

Mr. Soper. GEORGE A. SOPER, M. AM. SOC. C. E. (by letter).—It is by no means clear what sort of discussion was desired by the framers of this question. Its wording suggests that, in order to have clean water, we must either build filtration plants or purchase the land through which the water flows. This, of course, is not true.

In the average case, people using impounded supplies need not feel that they are driven upon either horn of this dilemma—certainly not to the purchase of whole water-sheds. This is at once the most expensive, least scientific, and most impracticable way to proceed in order to protect a water supply against pollution. For example, it would cost New York City about $58 000 000 to buy the drainage area of its new water supply, provided it could be had at the average rate of $100 per acre; and mere ownership is by no means an insurance against pollution.

The question suggests that pollution may perhaps be tolerated if only filtration is used. This is utterly wrong. It is wrong in

Mr. Soper. principle and not justified in practice. Filtration is a cure, and an excellent cure, as cures go, but it is not an infallible safeguard. Deliberately to allow filth to pollute a water supply with the idea that the water will eventually be cleaned by filtration shows a beautiful trust in filters and a greater confidence in the men who run them; but it is not good sanitary engineering. It has been well said that what we need in water is not repentance but innocence, and when it is possible to keep water pure this should be done by all means.

The extent to which water supplies can be protected against pollution depends upon a number of circumstances, but chiefly upon the amount of attention which engineers and others interested in pure water are willing to give to the subject. If waterworks people would give an amount of attention to regulating the drainage, inspecting the streams, and superintending the disposal of harmful matters on the water-shed commensurate with the time and money which they give to handling the water after they get it, nearly all danger from impounded supplies could be avoided.

Protection means a number of things which the public looks to engineers to understand, but which engineers frequently consider can be understood only by medical men; but physicians, as a rule, are not competent to handle sanitary matters. It is time that engineers accepted the responsibility, and undertook to deal with these subjects themselves.

The sanitary protection of a drainage area means, first, knowledge of the conditions where permanent and accidental defilement may take place; second, the careful design of measures to protect the water against these danger spots; third, adequate provision of law to carry out the measures; fourth, a settled intention to protect the water.

If the purpose exists, difficulties respecting the other requirements will disappear. Experience shows that filth can be kept out of water, if only there is a determination to keep it out.

Mr. Baldwin. WILLIAM J. BALDWIN, M. AM. SOC. C. E.—Some 25 or 30 years ago small filters were recognized as propagating beds. The speaker would like to ask whether there is anything in this, and what engineers know about it?

Mr. Williams. GARDNER S. WILLIAMS, M. AM. SOC. C. E.—It may be proper to say a word in regard to the kinds of bacteria that are to be dealt with, particularly in view of the statement of one of the speakers, that owing to the absence of light the number of bacteria may be very decidedly reduced in passing through a water-works system. It is to be borne in mind that those bacteria which are most dangerous to the human organism thrive in the darkness, to most of them light is germicidal, so that the above-cited condition would fail almost entirely to reduce the forms of bacteria which are in any way dangerous to health. Mr. Williams.

The subject of railroad pollution is certainly a serious one, as it can readily be seen that closing the closets while traversing watersheds might, in time, entail closing them altogether, and it seems that the more rational method of dealing with the question is either to provide receivers upon the cars for the dejecta, or to install such systems of drainage along the railroads as will divert the contamination from the water-shed.

It may be quite safely stated that while there is no question that a great deal can be done to improve a water supply by eliminating the sources of evident pollution from the water-shed, yet no such system will ever be perfect. It will never be possible to remove all such sources, and while the elimination of as many of them as possible is to be advocated, a community is not thereby warranted in using a surface water subject to possible pollution, without filtration. On the other hand, the mere filtration of the water or the mere existence of a filter operated by anybody is not a reliable safeguard. Whether the filter be of the slow-sand, or of the so-called rapid or mechanical type, it needs very close attention, and unless that careful watchfulness accompanies the operation, one cannot depend upon it any more than upon the ownership of a water-shed. A filter is a very delicate machine, and it must be carefully and skilfully handled.

It can be quite definitely asserted that, for a properly operated filter, there is no such thing as its becoming a propagation bed. The mechanical filter which is not washed at proper intervals will, of course, permit of the passage of bacteria by becoming filled, since the sand bed in such a filter operates simply as a strainer, and when the strainer is filled the material washes out at the bottom. In a slow-sand filter, however, there is no record known to the speaker of the condition suggested, in the 70 years' experience with that type; and with a mechanical filter, properly attended, nothing of the sort has ever occurred.

J. T. Noble Anderson, M. Am. Soc. C. E.—The speaker wishes to call attention to a point in Mr. Fuller's discussion, which must be due to some misapprehension on his part. He says that the water-sheds of several cities, among which he mentions Manchester, England, have been acquired at a cheap price. The speaker was employed many years ago on the Manchester water scheme, and also on that of Belfast, Ireland, and knows that the prices paid at both places were not what would be called cheap; for example, much of the Manchester watershed was acquired from the Fleming family, who asked $2 500 per acre for the land. The Manchester Corporation offered $100 per acre —the great difference is manifest—and it was purchased for from $400 to $500 per acre. Mr. Anderson.

Mr. Anderson. In the case of the Belfast water-works the average price originally paid for the land was about 10 shillings per acre per annum, which was the rental price paid by the farmer. This, with some other items, made the net return of the land required by the Belfast Corporation about $5 per acre per annum. With the compensation paid to the tenant and that paid to the landlord, the land cost the Belfast Water Board on an average of from $200 to $250 per acre.

There is another point, which has come up in Australia, which the speaker thinks has been generally overlooked by writers on hydraulic subjects, and that is, the use of the acquired water-sheds for re-foresting, especially in sub-tropical countries. The cutting down of forests interferes greatly with the rainfall, therefore it is not only important, in many countries, to protect the water-shed from impurity, but also to protect it from natural conditions, so that the rainfall of the range in which the water-shed is situated will not be impaired. By the judicious planting of these water-sheds that result is obtained, and the rainfall may be actually increased and a revenue earned. In the State of Victoria, the City of Melbourne has a water-shed roughly estimated at 100 sq. miles of forest, that will suffice to supply a population of 750 000, which it is expected will be reached within the next 8 or 10 years, and the Government has further reserved 250 000 acres all of which will be conserved ultimately and used by the Melbourne Water Commission. The point is an exceedingly important one on the subject in hand.

Mr. W. B. Fuller. WILLIAM B. FULLER, M. AM. SOC. C. E.—The speaker believes that all upland water supplies should be filtered, even though filtration is put in merely as a safeguard. Where it is suspected or known that pollution exists on an upland water-shed, the proper course to pursue is, first, to put in a filter, and then, if further funds are available, to buy up and remove the polluted portions of the water-shed.

While it is true, as has often been said, that "innocence is better than repentance," it must be recognized at the outset that no upland water supply, about which there could be any question regarding the necessity of buying up any portion of its area, can be pure; and unless the entire area is inaccessible to Man and beast, we have the ever present possibility of outside contamination of even the purest supplies.

Some of these possible sources of pollution have been pointed out, notably the contamination possible from railway trains traversing the water-shed and the need of keeping the closets closed during such times.

Passing over the difficulties and inconveniences attending the carrying out of such orders and the dangers present, supposing negligent attention on the part of the railroad officials, there are other sources of pollution of an inhabited region still more difficult to regulate, as, for instance, during the repair, construction, or reconstruction of railroads and roads where large numbers of laborers pollute the adjacent fields, the construction of country estates, the pollution by campers, picnickers and other nomadic hordes, and many other possibilities. Mr. W. B. Fuller.

It is true that large impounding reservoirs assist greatly in reducing the gross and accidental pollutions mentioned, but the standard of purity now demanded in many communities will not stop at one possible means of correction, but demands the best correction that engineering science can offer. To this end the best and safest guard against both accidental and permanent pollution is a properly designed and scientifically operated filter, the filtered water passing directly into covered conduits and not again exposed to the air, or possible contamination, until it reaches the tap of the consumer.

In advocating filtration the speaker does not wish to be understood as considering it a sure specific for all water evils, and he certainly advocates, first, the securing of the purest water supply from the least polluted upland possible; and, second, the keeping of this upland supply as pure as possible before delivery to the filters. To this end, he is a firm believer in the necessity of draining all stagnant and marshy places; the removal, by purchase or otherwise, of undesirable persons and manufacturers from the water-shed; the organization and efficient maintenance of sanitary patrol; and the taking of all possible precautions against pollution. The less the filter has to do, the safer the water, but the filter should stand at the outpost and guard against enemies that slip in unawares, as well as against those that are known.

J. WALDO SMITH, M. AM. SOC. C. E.—A large impounding reservoir at the lower limit of a water-shed is, from a sanitary point of view, one of the greatest safeguards which can be provided for a water supply. The sanitary quality of a water is undoubtedly increased by long periods of storage, though the storage may have but little effect on either its taste or appearance. In fact, the physical quality of a water may be made worse by storage, and for this reason, as well as to satisfy the proper public demand for a water, clear, and pleasant to look at, filtration is necessary. Mr. J. W. Smith.

The question as to whether or not conditions in a storage reservoir can be improved by stripping its bottom and sides of most of the original organic matter is an important one. Recent studies indicate that the effect of stripping is to guarantee a somewhat better quality of water for a few years, but that, in the end, the conditions will be the same as though but little work in this direction had been done. An unstripped reservoir will improve; a stripped reservoir will grow worse until finally, if all other conditions are the same, the quality of water delivered by two such reservoirs will be identical. Even at its very best, stripping is not an absolute guaranty; vegetable growths may occur in any reservoir and can only be guarded against by thorough aeration followed by filtration.

Señores Solís and Guzmán.

Señores Pablo Solís and Octavio Guzmán* (by letter).—The writers congratulate themselves, as representatives of the State of Puebla, upon the opportunity that now presents itself of knowing, through this meeting of eminent professional men, the present state of the art in the five technical subjects which make up the programme. They regret that this programme lies without the field of professional practice of their State, which is purely textile, and in whose capital, on account of its excellent climatic and topographic conditions, it has not been necessary, up to the present time, to introduce modern sanitary devices. On account of the extraordinary firmness of its soil, it has not been obliged to contend with exceptional foundation problems; and, as the center of an extensive network of railroads which has completely done away with care of the ancient roads, it has not preoccupied itself concerning the future fate of the public highways. The recent growth of automobiling, however, now compels attention to, and interest in, them. On the other hand, our capital being on the eve of establishing a system of drainage and all manner of sanitary improvements, the subject of potable water supplies also greatly interests the State, and, through the writers, it will secure, from the deliberations of this meeting, knowledge of the latest advances in all the important matters proposed for discussion, and the writers will endeavor to co-operate by discussing the two first questions of the programme presented, giving their opinions on these subjects on an abstract basis.

With regard to the first subject under discussion, the writers believe that, with the present state of knowledge and experience in the art of sanitation, only the second alternative stated in the question is practical, that is, to permit a certain amount of contamination, and to rely on a subsequent purification.

In truth, the art of sanitation, in spite of the great advance that has been made in the acquisition of knowledge and of practice, has, up to the present time, only undertaken the sanitation of isolated and independent communities. In sanitary matters facts must be considered in their multiple condition, not isolating some with respect to others, because they are never independent. From this point of view it is of little use to attend to the sanitation of one city, if the valley or region in which it lies is not also taken care of, and in the sanitation of any region the disinterested co-operation of the various communities in it is indispensable.

Unfortunately, up to the present time, each community watches its own interests only, not concerning itself with the interests of the others with which, on account of local conditions, it is inseparably linked, and on whose healthfulness depends in great part its own. If

* Delegates, to the Convention of the American Society of Civil Engineers from the State of Puebla, Mexico. (Discussion translated from the Spanish.)

Señores Solís and Guzmán.

a community does not dispose properly of its refuse, getting rid of it without special treatment of any kind, its final disposition being near neighboring communities, the latter will become contaminated, and disease will be developed in them which may infect the culpable community and even the whole region. Therefore, sanitation, to be complete and effective, must be more than local, it must cover the entire region, and must embrace all towns that are linked by natural conditions; and as long as this desideratum of the sanitary art is not reached, the potable waters in such regions, whatever their origin may be, if they are below the level of the sewage of other towns, will be somewhat contaminated. Even granting great vigilance and control by the Governmental powers, the result will be the same, because unless such regulation is mechanical and automatic, it will be inefficacious. It is fruitless, therefore, to purchase and watch the water-sheds that feed the reservoirs, and, in the opinion of the writers, it is best to allow contamination to take place to a certain degree, trusting to filtration for purification. The question may well be asked: Is regional sanitation which will prevent the contamination of water possible in practice? Judged purely in the abstract, and by the theories of sanitary art only, an affirmative answer forces itself upon us, and, furthermore, that this is the only proper form. It is incomprehensible that any system which restricts its attention to isolated points can be called sanitary. The phenomena of life, under whatever form we may consider them, be it individually or collectively, are always the same; and as, in the individual, health cannot exist unless it exists in the whole organism, in the same way, collectively, it will not exist if it does not encompass the whole region. Now, what are the fundamental conditions necessary to regional sanitation, in the strictest conception of the term? The very same conditions as are necessary in individual life; and, summed up, they may be condensed as follows: Continuous renovation; continuous renovation by circulation or motion; and continuous renovation by chemical or substantial transformation. In the life of the individual these conditions in their admirable harmony are regulated by what is known as "the vital principle," and in collective life they are, in turn, regulated by the laws of political economy, and the rights of the individual under the direction of the Government. It is within the field of political economy to arrange financially for the problem of continuous circulation by chemical or substantial transformation, and it is the arduous task of the Government to harmonize sanitary interests in such a way that the best interests of the whole region will be preserved.

In the same degree as these difficulties are overcome, will the art of sanitation advance from being purely local, as it now is, to being regional, as it will be to-morrow, and in that same degree potable waters may be considered as safe from contamination.

Señores Solís and Guzmán.

In what degree has the scientific disposal of garbage been successfully and economically applied to the sanitation of towns? Much effort has been made in this direction, and results have been obtained, propitious in some cases, adverse in others, but thus far the advantageous results have not been confirmed by properly made estimates, and the general opinion among sanitary engineers is that we have not arrived at satisfactory results, especially with respect to sanitation by gravity, in which process the matter being highly dissolved, its extraction is very costly.

Now while the art of sanitation remains in the precarious condition of not being able to get rid of detrimental matter except by the process of solution and removal, the contamination of potable waters is inevitable, and it is necessary to resort to filtration, which is the only solution of the situation in the present state of the art. But when the economical transformation of domestic and other wastes becomes possible, to the extent that it may be applied generally, and when the mental condition, customs, and altruism of the community have advanced to the perfection required by the present state of sanitary science, it will be possible to put its teachings into practice. The decisions of Boards of Health may then be carried out with efficiency, and then regional sanitation may be accomplished, and the non-contamination of potable waters will also be an established fact. In the meantime we must be resigned to allow potable waters to become contaminated, and we must purify them by filtration.

Mr. Russell.

S. Bent Russell, M. Am. Soc. C. E.—It is evident from this discussion that the question as to which procedure is the better policy cannot be met with an answer so simple and direct that it will not need much explanation and limitation as to conditions.

Mr. Fuller has set forth in an admirable way the facts which enter into the problem. He has weighed them in an impartial spirit, and his conclusions, for the most part, seem to be logically drawn. To the speaker, however, it would seem that part of his first conclusion has not been fully demonstrated in his argument. It may yet be shown that, with equal intelligence and watchfulness, the purchase and control plan is as efficient as filtration, with regard to the sanitary character of the water. The cases referred to by Mr. Snow will bear this out.

The other part of this conclusion—that in regard to the physical condition of the water—is undoubtedly sound, and is a very strong point in favor of filtration. This advantage will usually outweigh the detrimental factor of sentiment against impure water artificially made pure. The truth of this has gained more general credence with the growth of the science of water purification.

To make the involved point more clear, compare two towns, one with a water-shed owned and controlled, the other where a certain amount of pollution is allowed, but filtration is in use. Which has the better water supply? This is still a debatable point among engineers; some favor the former, others as strongly the latter. Either plan will give reasonable safety from water-borne diseases, provided intelligent care is exercised in the management. It might be said, as in the matter of forms of government: Mr. Russell.

"* * * let fools contest;
Whate'er is best administer'd is best."

With good administration, the former town will have the better supply, from the point of view of one who thinks a naturally pure water is better than a purified water; while the latter town will be better protected usually against objectionable tastes and odors and visible impurities, as shown by Mr. Fuller. It may be remarked that a majority of water consumers do not inquire into the purity of the supply, if the water is clear and sweet. On one side, therefore, it is found to be a matter of sentiment or prejudice in favor of water naturally free from sewage pollution; on the other, that of prejudice against water which is at times objectionable to the senses. The speaker believes that of late years progress has been in favor of filtration.

If it is admitted that there is no great balance of advantage in favor of either plan as regards quality, the question of cost may then be taken up. At the final test a community usually wants a satisfactory water at the lowest cost. Any further expenditure must be more than justified, in money value, by the superior quality. It must be recognized, as Mr. Clarke says, that the financial question is often a controlling factor. Mr. Fuller shows that, in the matter of cost, filtration has the preference. In any given case, however, the cost will depend upon local conditions, and, therefore, the procedure must be governed by these conditions. Among the local conditions, existing laws and customs, as affecting sanitary control, must be considered.

Mr. Fuller states that there is need for better legislation regarding sanitary matters, and Mr. Snow shows most clearly what proper legislation can do for the protection of water supplies. This brings us to the broad question of proper legislation for controlling the disposal of sewage and wastes, and the pollution of lakes and streams. The speaker understands that, in the matter of such legislation, many of the States, and the United States, as a whole, are sadly behind the best practice of European countries.

These, of course, as far as pertains to the United States, are all questions of the future. Engineers are now working toward the adoption by the Government of wise laws for the protection of future generations against the dangers that will come with increased population, and they should continue in this work with still greater zeal and

Mr. Russell. co-operation. Such laws are required by true public economy in the matters of sanitation and water supply. Proper legislation will reduce to a minimum the expense of protecting the water-sheds.

From these considerations, with a view to future public economy and health, the following may be formulated as an answer to the question under discussion: The best policy is to limit the pollution of the water-shed by suitable legislation leading to proper drainage and disposal of wastes, followed, of course, by vigilant enforcement of the laws, so that they will effect their purpose. This should be supplemented by filtration of such character and efficacy as to produce a satisfactory public water supply, as defined by Mr. Fuller.

In the speaker's judgment, the cases requiring purchase of land instead of filtration would be so few that they could be classed as exceptional.

Whether or not this answer is accepted, all may fairly hope for, and work together for, legislation to protect our streams and lakes from pollution, and for the better observance of sanitary laws by the public.

Mr. Sherrerd. M. R. SHERRERD, M. AM. SOC. C. E.—It seems as if there might be a middle ground between the views expressed in the discussions by Mr. G. W. Fuller and Mr. George A. Soper, one of which points to filtration as the only means of safety, while the other seems to indicate that filtration is not necessary. It seems to the speaker that proper sanitary protection turns on the point of whether water-sheds can be selected which will be safe without filtration; perhaps it will be essential that, in every case, filtration must be resorted to eventually. It would seem possible, where the population is small and perhaps occupies only a portion of the water-shed, that the installation of a sewer system to take off the waste from that population, which represents the more expensive portion of the water-shed (from the standpoint of purchase), could be resorted to, and that the outskirts, which are generally the cheaper part, should be purchased, so that a combination of sanitary measures would be effective for a great many years. This would certainly be a cheaper method in many cities in the United States than the immediate installation of filtration, and would be also a very satisfactory protection for many years.

One of the discussions referred to the effect of sedimentation, and it might be of interest to give some figures as to the reduction of bacteria noticed in the water supply designed for a population of some 300 000, where the capacity is about 50 000 000 gal. daily. On this plant there was recently constructed a storage reservoir near the city holding about a month's supply. Formerly, the water was brought down through some 25 miles of pipe line, and, at the intake, the bacteria would run as high as 5 000 per cu. cm., and in the tap water of the city it would be reduced, principally due to the absence of light

Mr. Sherrerd.

in the pipe lines and sedimentation in the city reservoirs, to an average of 200 bacteria per cu. cm. After the use of the storage reservoirs, where the water is held for about a month, the bacteria being about the same at the point of intake on the water-shed and the conditions being similar, the number of bacteria in the tap water has been reduced to an average of from 20 to 30 per cu. cm., running along at those figures for months at a time.

There is one other feature in the protection of water supplies which seems to be essential and to which more attention will no doubt be paid in the future; that is, the danger of pollution from railroads traversing the water-sheds. It would seem conclusive, and public health authorities have recommended, that stringent laws will be necessary to secure proper sanitation in this respect unless the railroads see that it is to their interest to provide some sort of sanitary measures. In New Jersey it was proposed that they should go even so far as to enact legislation which would put this matter in the hands of the State Board of Health. This was done at the suggestion of some of the local authorities who were interested in the matter, and at least one railroad has agreed to close the closets on the trains while traversing the water-shed of the Newark water plant. This is done also on the New York, New Haven and Hartford Railroad on the New Haven water-shed, and is being done on the water-shed of the Scranton, Pa., water supply because of an order from the State Board of Health. It was also proposed to do this at Seattle, Wash., on the portion of the road which was to be built through the territory from which that city derives its water supply.

New Jersey has enacted legislation which will allow the cities to provide sewerage facilities for the drainage of the population on water-sheds where the State Board of Health creates a district, and the State Sewer Commission approves the plans. The cities are to pay for the installation of these systems.

To insure that the sewage be treated properly before it is discharged into the stream, it is proposed, on the plan to which reference was made, that a sewer system shall be installed with septic tanks, and that broken-stone sprinkler filters and fine sand filtration shall be resorted to before the outfall discharges into the main stream below the intake. This act provides that, while the city shall pay the entire cost of installation, it may have the authority to enter on private property and cause connections to be made from outhouses or closets to the sewer, and provide proper sanitary arrangements; and it is also proposed that a water supply shall be installed from one of the reservoirs in order that the sewer may be flushed properly.

In the speaker's judgment, the purity of the water supply for this particular city will be insured, at least for a great many years. There is one question, however, which is troublesome, namely the fact that

Mr. Sherrerd. by the installation of such a system, particularly in the neighborhood of a suburban section of the metropolitan district around New York, the property, to a certain extent, is made more desirable for residence purposes, particularly for villa sites where people from the metropolis may have all those advantages, so that this has, in a measure, a tendency to increase the population of the district. On the other hand, to attempt to secure by purchase the tracts of land that are necessary to protect a water supply is usually much more expensive than the cost of installing and maintaining adequate sewerage facilities.

The conclusion of the speaker is that conditions must always govern, and that it is essential that the purest water should be obtained primarily; its treatment will depend on circumstances.

Mr. Blackford. F. W. Blackford, M. Am. Soc. C. E.—This question is largely one of economics, and should be considered like any other problem that comes before an engineer. Local conditions make a great difference in such propositions. In populous districts, like those adjoining New York, and in Pennsylvania or New Jersey, and other places similarly situated, it might be impracticable to acquire a water-shed entirely. but in many places it would be practical. Of course, this depends largely on the ability of the community to tax itself sufficiently to obtain a large water-shed or one of sufficient size to furnish a supply, but it is a question of locality and a question of making figures, like other propositions, such as building railroads and adjusting the grades and curves to the amount of money available, and to the traffic which is expected. In many places, even in populous districts where the water supply is drawn from a distant point, a large tract of land could be acquired and controlled absolutely. It does not follow that such land should remain useless; it could be planted in grass or in forest—more properly in forest, which, in the course of a generation, would become profitable; and, while it might also be found necessary to filter the water to some extent, the circumstances might be such that it would not have to be filtered at all. It is an economic proposition, and one should consider the amount which the community under the law could tax itself, the amount of income to be derived, the operating expenses, etc., as against the fixed charges.

Mr. Johnson. George A. Johnson, Assoc. M. Am. Soc. C. E. (by letter).—It is no simple matter to make a sizable water-shed hygienically clean and keep it in a thoroughly safe condition, either by purchase, depopulation, or sanitary patrol. The cleanest and most carefully policed water-shed is likely to receive "accidental" pollution, at one time or another, through the discharge upon it, or into the streams draining it, of typhoid-infected excreta by a chance traveler suffering from a mild case of this disease, that is, "walking typhoid." This thought as to danger in unsought places may be carried even further by directing attention to the fact that typhoid bacilli may persist in the urine of convalescent typhoid patients for weeks and even months. Mr. Johnson. Some pathologists claim to have detected this organism in the urine several years after convalescence.

It is a fact that, in general terms, the degree of danger existing in a water supply may be measured by the cleanliness of the water-shed; in a sense, it stands in inverse proportion to it, other things being equal. There can be no doubt that an impounded supply which is filtered will always be safer than a filtered supply which is not impounded, owing to the natural purification effected in impounding reservoirs.

River waters which are derived from populated water-sheds, no matter how efficient are the methods of disposal of the wastes of the resident population, cannot be looked upon as safe in any sense, if such waters are pumped direct from the river and supplied to the consumers in an unfiltered state. Such waters may be made pure, according to existing standards, by filtration, but the degree of purity of the filtered water depends much upon the nature of the filtration process, the suitability of the plant, as built, to meet existing conditions, and, finally, but by no means least, upon the manner in which the filters are operated.

It is well known that polluted surface waters which are impounded in large reservoirs undergo a material purification due to natural agencies, the extent of such purification being greatest where the period of storage is longest. Published data show that at Washington, D. C., where the period of stay of the water in the series of reservoirs ranges from two to four days, the bacterial removal amounts to about 93 per cent. A similar removal was effected in the reservoir at Lawrence, Mass., when the average period of storage of the raw water was about two weeks. Equally good, and doubtless better, results may be generally expected in the case of large impounding reservoirs, such as the Wachusett, Croton, and Boonton, where the average storage periods have been roughly computed as about 532, 235 and 244 days, respectively. Bacterial deposition in fairly clear water is slower and generally less in amount than in very muddy water.

Such information as is available indicates clearly that typhoid germs do not multiply in water. In the course of a week or two a large majority apparently dies out, due to such causes as osmosis, sedimentation, unfavorable environment, lack of a suitable food supply, antagonism on the part of other bacteria, and to the fact that all bacteria serve as food for some of the higher organisms. It appears to be certain that a small percentage of the typhoid germs, constituting the more hardy forms, will live in water and mud for an almost indefinite period. The virility or pathogenicity of these resistant forms is something about which but little is definitely known. Whether or not their ability to produce typhoid fever is equal to the average

Mr. Johnson. pathogenicity of all the typhoid germs at the time of their discharge into water is not certain. It probably approaches such a standard as indicated by the quite lengthy list of epidemics following spring thaws, and the washing into streams of long-standing infected excreta.

Mr. G. W. Fuller. GEORGE W. FULLER, M. AM. SOC. C. E. (by letter).—The discussions on this topic appear to cover the question quite fully, as it is now understood. Probably the most gratifying feature is the almost unanimous viewpoint that a better quality of water is desired than formerly, and that, to this end, there is no doubt that better laws are required. The State of Pennsylvania has recently taken a long step forward, as described by Mr. Snow. It is also clear that in future sanitary developments, as pointed out by Dr. Soper, the engineer is likely to assume more responsibility, and to direct various details, in keeping water-sheds clean, to a greater extent than has been generally the case hitherto.

Local problems can obviously be best solved only after weighing the local evidence. In general, preference now appears to be given to that treatment which will afford satisfactory results at the least cost.

On this basis, it seems clear to the writer that there is bound to be rapid progress made in the introduction of filters for ordinary impounded water supplies, even where the storage period is comparatively great, and also in the systematic elimination of constant sources of pollution as well as in safeguarding against so-called accidental pollution.

The main difference in opinion seems to relate to a choice of the two treatments under consideration for places where the cost may be about the same. Sentimentally, the preference in the minds of many lies with the unpolluted source, and there is much to substantiate that viewpoint, especially for small water supplies. For large water supplies, the writer believes that the pollution should be eliminated as far as practicable, without the purchase of much land, and filtration should be generally adopted, both for the sake of the hygienic quality of the water and also for its general cleanliness, as mentioned by Messrs. Smith and W. B. Fuller.

It seems futile to say more about the relative merits of these two problems until further detailed evidence is available. In preparing his earlier statements, the writer gave more weight than perhaps is indicated to the belief that the filtration of so-called pure waters will produce improvements that perhaps some of the discussors do not realize. The Little Falls filter plant, treating a fairly safe upland water, has certainly brought about a supply, not only of attractive appearance, but also a surprising reduction in the prevalence of typhoid fever in Paterson and Passaic, N. J. This is in line with numerous experiences abroad. It is also the view of Health Commissioner Darlington, of New York City, with regard to the filtration of

Mr. G. W. Fuller.

the Croton supply which, for years, was considered of a good quality, according to the standards of earlier years. After the citizens of New York have an opportunity to enjoy the benefits of the recently recommended filter plant for the Croton supply, the writer is positive that the benefits will be considered as worth far more than their cost.

As to the inquiries made in the course of the discussion of this topic, it is to be stated that the propagation of bacteria, as mentioned by Mr. Baldwin, is not an unusual occurrence in small house filters which are infrequently and improperly cleaned. As stated by Mr. Williams, bacterial growths are quite unusual in large filter plants which are well built and well operated. They have occurred at times, however, both in test filters and in large filters in actual practice. During the summer of 1892 bacterial growths developed in the experimental sand filters at the Lawrence Experiment Station, as described by the writer in the 1892 Report of the Massachusetts State Board of Health. The reports to the Local Government Board of England, by the late Sir Edward Frankland, show that bacterial growths developed in some of the London sand filters during the winter of 1894-95. Bacterial growths were found in the experimental mechanical filters at New Orleans, as stated in the report of R. S. Weston, Assoc. M. Am. Soc. C. E., upon his investigations made in 1901.

Mr. Anderson refers to the land costs of the water-sheds of British cities: These costs do not seem to the writer to be very high, but they are much in excess of recent figures, stated to him as paid for other and presumably more lately acquired water-sheds.

Mr. Sherrerd's references to the experience of Newark are all of interest and importance, and emphasize the fact that local conditions have to be carefully weighed. Without doubt, large storage reservoirs away from the water-shed and the purification of all sewage on the sparsely populated water-shed are helpful, as is also true of the natural aeration of the water drawn from near the bottom of the Newark impounding reservoirs. In the absence of evidence of such water being prejudicial to health, and of being objectionable to the senses, it is not likely that large existing impounded supplies of that type will be generally filtered until more evidence is obtained to show the benefits of filtration as applied to such cases, as regards the improvements effected in the physical character of the water and as to its sanitary quality.

George W. Fuller

George W. Fuller was a product of the Massachusetts Institute of Technology (B.S. 1890) and the University of Berlin. The first four years of his professional life were at the Lawrence Experiment Station (Massachusetts) where he achieved major success in his research on water purification and sewage treatment. In 1895 he undertook experiments at Louisville, Kentucky and later at Cincinnati, Ohio on rapid sand filtration techniques. By 1899 his reputation enabled him to establish an independent practice. Two years later he joined Rudolph Hering in a partnership which continued for a decade. Later, in 1916, he joined James Robinson McClintock in a partnership which lasted until Mr. McClintock died in 1933. For thirty-four years Fuller enjoyed success as an international consultant in sanitary engineering. He served more than 150 cities from Washington, D.C. to Shanghai, China. In many projects he was given sole responsibility.

Among his more famous projects was the Ashokan Reservoir (NY) and the Wards Island (NY) activated sludge plant.

George W. Fuller was a prolific writer contributing thirty papers to the American Society of Civil Engineers alone on such diverse topics as a National Reclamation Policy (1930), Chmi-Hydrometry (1916), and Disposal of Municipal Wastes (1905).

Vol. XLII. DECEMBER, 1899.

AMERICAN SOCIETY OF CIVIL ENGINEERS.

INSTITUTED 1852.

TRANSACTIONS.

No. 857.

STREAM CONTAMINATION AND SEWAGE PURIFICATION.

An Informal Discussion at the Annual Convention, June 28th, 1899.

SUBJECTS FOR DISCUSSION:

(*a*) "Should Stream Contamination by the Sewage of Cities be absolutely prohibited by law?"

(*b*) "Should the Purification of the Sewage of Cities be Compulsory; and is this feasible for Large Cities?"

(*c*) "Is Filtration the coming solution of the Pure-Water Question for Cities?"

By Messrs. R. E. McMATH, W. C. PARMLEY, GARDNER S. WILLIAMS, PETER MILNE, KENNETH ALLEN, JAMES OWEN, CHARLES G. DARRACH, E. W. HARRISON, P. A. MAIGNEN, L. L. TRIBUS, PALMER C. RICKETTS, JAMES S. HARING and JAMES H. FUERTES.

Mr. McMath. R. E. McMATH, M. Am. Soc. C. E.—This question has been subdivided into three topics, the first being: "Should Stream Contamination by the Sewage of Cities be absolutely prohibited by law ?"

This question may be answered in the affirmative on the ground of the common-law principle: "Each must so use his own as not to injure any other." No individual can discharge filth or other objectionable matter into a ditch or drain which traverses the field of his neighbor. If he attempts to do so the court will afford the injured party a remedy. If a city, town or village defiles a stream with sewage, the law is an already existing remedy, and has been applied in many cases which could be cited. Mr. McMath.

The presentation of this topic for discussion by this Society does not imply any doubt as to this recognized law, but, perhaps, is intended to call the attention of engineers to its existence and to the necessity for its extension to questions which, in this country, may arise between the people of adjacent States, if the defiled stream is a boundary, or passes from the territory of one State into that of another. That is to say, is there not a necessity for a general law which will embrace all States, and will settle all questions which arise between States?

Several of the States do exert their powers to prevent contamination of streams by the cities under their jurisdiction, but there is, as yet, no statutory provision to protect against contamination by the sewage of cities in other States. If, by the indifferent or mistaken policy of one State, its cities, great or small, are allowed to defile the streams, must the people of other States suffer for lack of adequate remedy?

The established law referred to has not been enforced in many cases where the transgressor is a city, but the transgression being for a time condoned does not establish a right to transgress for all time. As the country increases in population, the pollution of streams will increase and will quickly pass beyond possible toleration.

As the speaker views the matter, a law more imperative than any statute should control every engineer who has to do with the sewerage plans of cities; and he should foresee and provide for the disposal of sewage eventually, if not immediately, in another manner than by discharge in crude form into streams. Anything short of this is not worthy to be called a plan for a sewerage system.

The City of Chicago has for years been preparing a channel for the discharge of its sewage through its back door, so that this sewage will pass the front doors of other cities, and some of these other cities are not well pleased by the prospect, especially since the demonstration that certain diseases are water-working and that the pathogenic bacteria are tenacious of life. The topic under discussion would be an important one if Chicago did not exist, but the Chicago case gives much emphasis to the topic at the present time, for it is a scheme to do permanently, and on a large scale, what everyone must admit should not be done at all—a scheme which is to-day without an avowed designer and almost without a defender.

There was a time when many believed that the water of a river purified itself. Others firmly believed that dilution to a point beyond the reach of refined chemistry must render such water innocuous. The Chicago scheme had its origin when these beliefs were held. This scheme, involving the expenditure of more than $30 000 000, has progressed, notwithstanding the passing of the faith upon which it was founded, and now the only plea which is advanced for its consummation

Mr. McMath. is the fact that $28 000 000 and more have been expended on it and it is now too late for objectors to state their objections.

This brings up the consideration of the second topic: "Should the Purification of the Sewage of Cities be Compulsory; and is this feasible for Large Cities ?"

The answer to the first part of this proposition follows that given for the preceding topic. If streams are not to be contaminated, the sewage must be purified; for water is the only known vehicle which can carry the wastes of human life and industry away from our homes and shops. That such can be purified, so that the effluent may be entirely unobjectionable, has been demonstrated beyond occasion for further discussion. How to do it, in each case that may arise, is the work of engineers, to plan, execute and operate.

There is no casuistry in the assertion that in this matter the end justifies the means, even when the difficulties surrounding the end require large and many means.

As to its feasibility for large cities there are too many old-world examples of successful solution for American engineers to hesitate as to the answer. It can be done, and we are the men who can do it.

The third topic comes in to round out the discussion: "Is Filtration the coming solution of the Pure-Water Question for Cities ?"

Asked in connection with the preceding topic, the question may be taken as a suggestion that the sufferer from sewage contamination of water supply must protect himself, and thus relieve his up-stream neighbor of his obligation.

To meet this suggestion it is sufficient to mention the fact that the sufferers do not all live in cities, and are not able to protect themselves. There are evils from such contamination not measurable by the results of a bacteriological examination for specific microbes. There is also a contamination of water by the drainage of lands outside of cities which sooner or later will justify and even compel resort to filtration of the water supply of cities.

The doctors may say that they can cure a large percentage of cases of diphtheria, for instance, but the wise mother does not expose her children to infection on the chance that all cases in her family will be cured.

Therefore, filtration may promise a remedy for defiled waters, but that does not imply that the filtered water would not be better and safer had it not been contaminated.

Mr. Parmley. W. C. Parmley, M. Am. Soc. C. E.—The question of stream pollution is one that concerns almost every city. If no pollution of the water courses is to be permitted, then all the sewage must be purified, and the question resolves itself into one of expense. Small towns, or those of larger size, which are favorably situated, may adopt some method of ultimate treatment at comparatively small cost, but in the case of many, if not indeed most, of the larger cities, the expense involved by such treatment would be a very great burden, if not altogether out of the question. Insistence upon absolute freedom from pollution would make the interception of sewage and the overflow of storm water an impossibility. Mr. Parmley.

The whole matter is one in which the ideal is set over and against the practicable—the desirable but extremely difficult to attain, against the less desirable but practically possible—but it is no more important that all the water we drink should be free from impurities than that the air we breathe and the food we eat should be pure and wholesome. To insist, however, upon a high degree of purity in a stream from which no drinking water is drawn, or at most only stock water, is to require heavy payment for very small benefits, and many streams have long since ceased to furnish even the latter supply.

The only other cases necessary to consider are those in which the odors arising from a polluted stream become unpleasant or harmful, or where the sight of turbid or colored waters or the refuse left by them along their banks is objectionable for æsthetic or sentimental reasons. The sanitary welfare of the people within the area affected is the ultimate and only condition to be insisted upon.

Even in the earlier development of our cities the natural drainage courses soon become so polluted by local discharge that no drinking supplies are likely to be taken from them, and it remains to be determined whether the pollution which has already occurred shall be removed or not. As most of this pollution is from factory wastes or from the washings of stables and privy vaults, it is a source of impurities nearly impossible to avoid. While the drainage area remains suburban, the cost of adequate sewerage would be too expensive to be borne by a sparse population or an isolated manufacturing company, and when the density of population becomes sufficient to make thorough sewerage possible, the soil and water-courses have already become so polluted that many years would be required to thoroughly drain them. Even then, as those who have had experience will testify, it is impossible to prevent entirely the clandestine discharge of some sewage into the streams. It follows, then, that the smaller courses which are not entirely taken into the sewerage system will have a considerable degree of pollution. Just the percentage which may be allowed is impossible to state, but it is dependent on the character of the polluting matter, the climatic and many other conditions. It is safe to say, however, that no stream should be permitted to become so befouled as to give off offensive effluvia and gases, thus showing that it contains a large amount of putrefying matter. Either, these streams should be diverted into the sewers, or the whole stream taken up in a large trunk sewer.

The question of permitting storm water from sewers to overflow into

Mr. Parmley. water-courses is important in many cities. If such overflows are not allowed, the cost of the sewers thus required, over that where the overflow system is permitted, becomes an item of no mean proportions. The engineering department of the City of Cleveland has at the present time a contention with the Board of Park Commissioners regarding the overflow of storm water from a sewer into a stream; and if it is finally settled that the sewer cannot be permitted to overflow, the cost of the sewer will be about $250 000 more than if it were constructed with storm-water overflows. In another instance the local court has issued a restraining order to prevent the discharge of excess storm-water into a stream which is already highly polluted. The final settlement of the case is of great importance, as upon this decision will depend other important projects in the sewerage of that city. The court apparently ignored the important facts brought out in the trial, viz., that this overflow would not operate except in case of excessively heavy storms, when the water would be discharged for perhaps 15 or 20 minutes, that these overflows would probably occur only a few times during the season, that the actual stream in time of flood would be nearly as impure as the water overflowed from the sewer, and that even if it were not, the time of maximum flood in the stream would be after the sewer had ceased to overflow and the stream would continue to flow with a large volume for many hours afterward, thus washing away the last traces of any fancied pollution.

If sewers are not allowed to discharge or overflow into natural water-courses, in many instances the double system of sewerage will become an absolute necessity. If the sanitary sewage be conveyed to a proper outlet through separate sewers and all the storm-water is to flow to the natural drainage channels, either through storm sewers or directly over the ground, the water discharged therefrom will usually be more foul than that which reaches them from the storm overflow structures of the combined sewers. As all engineers have probably noticed, the first washings of the street surfaces are extremely foul, and, by the system of overflows, this foul water is carried to the sewer outlet while a portion of the cleaner washings which occur later in the storm, mixed with the normal volume of sewage, will escape from the overflows into the stream.

If, however, the contaminating effect of street water be neglected and only the effect of sewage from the sanitary flow be considered, and if the entire volume of dry-weather flow in the sewer be defined as sewage, in the first case above cited it was shown that the maximum degree of pollution, that is, when the sewer is just at the point of overflowing, is 1 of sewage to about 7 of street water, and when the structure is discharging its maximum volume of overflow the proportion is 1 of sewage to 30 of street water. Now, consider the degree of pollution of the stream into which this discharge is made, and the proportion of pollution is much less in proportion to the volume of water already flowing therein. Mr. Parmley.

It is a matter of the simplest observation to show that in many cases the water escaping from overflows in time of storm is not more foul than that in the swollen waters of the flowing stream. It being thus impossible to prevent the pollution of small courses flowing through thickly populated areas, it seems unwise to attempt to preserve in them a greater degree of purity than that required to prevent their contaminating the surrounding air, and when they are so short, or their velocity of flow is so rapid, that an outlet is reached at a point beyond which it is unnecessary to consider their impurities, and before those impurities have had an opportunity to decompose, the further expenditure of money, in order to preserve a high degree of purity, would be a needless waste.

The matter then suggests the following conditions, with their appropriate treatment:

1st.—Streams or bodies of water from which potable supplies are drawn should be preserved in the purity required by the most advanced sanitary science, and if such purity can be obtained in no other way, the supply should be properly purified before being used.

2d.—Streams or ponds which have sluggish motion with consequent tendency to stagnation should be kept as free as possible from sewage pollution in order that their emanations may be less dangerous.

3d.—Streams in which the flow is so rapid, or in which the courses are so short, that stagnation and decomposition are impossible within the limits of time necessary to consider, may be used advantageously for the overflow of excess storm water, or as the direct outlet for a limited number of sewers.

Gardner S. Williams, Assoc. M. Am. Soc. C. E.—It would be interesting to determine, if possible, where to draw the line as to the limit of the stream which is not used as a source of potable water. Nearly all streams flow into others or into lakes, and at some point drinking water is being taken from them. Mr. Williams.

An interesting case came under the speaker's observation, while he was connected with the water supply of Detroit, Mich., which involved the investigation of a serious typhoid epidemic in that city in 1892.

Port Huron is situated at the head of the St. Clair River and at the foot of Lake Huron. The distance by water from Port Huron to Detroit is about 60 miles, about 18 miles of which is through Lake St. Clair, where the current is very slow, and which acts as a monstrous settling basin, giving ideal conditions for sedimentation and the consequent purification of the water. The sewage of Port Huron is discharged into Black River, a very sluggish stream, which flows through the center of the city and receives sewage from both sides. The current in Black River is practically nothing, except in time of

Mr. Williams. freshet. No water for potable purposes is taken from this stream, but the Detroit and St. Clair Rivers and Lake St. Clair are sources of supply.

In 1890, Congress decided to improve the navigation of Black River, and work thereon began in 1891. Dredging was commenced at the bar at the mouth of the river, and was approaching that part of the river where the principal sewers discharge when cold weather stopped the work. On April 16th, 1892, work was recommenced at a point about 300 ft. below the first bridge, and at that bridge two large sewers discharge into the river. Below that bridge it is reported that a depth of 12 ft. was dredged through an accumulation of sewage deposits which had been gathering there for years. The inspector's records of the dredging show that there were excavations in such material 8 ft. in depth. This material was put in dump scows, taken out into the St. Clair River and dumped at a point over the Grand Trunk Railroad tunnel, or approximately 60 miles above the intakes of the Detroit Water-Works, which are just below Lake St. Clair, and virtually at the head of the Detroit River.

The record of typhoid fever in Detroit at that time is interesting. There had been some deaths from typhoid early in the year, but between May 11th and May 28th, 1892, no deaths were reported. On May 28th there was one death. On June 5th there were four, and during the ensuing 25 days there were thirty-seven deaths from typhoid fever, the population of the city being then about 230 000. The death rate from typhoid in that year in Detroit rose as high as the typhoid death rate of any large American city, except Chicago, Pittsburg, Jersey City and St. Louis, in the past ten years. There were three years in which the death rate in Pittsburg and two in which that of Chicago had been higher.

It was somewhat difficult to trace the cause of this epidemic, and when the speaker began his investigations he was not aware of the operations of the Government in Black River in 1891 and 1892. He therefore did not look there for the pollution, but began examining the river and studying rainfall, the effect of thaws, the effect of ice in the river and the character of the water. None of the ordinary methods of accounting for the presence of typhoid was applicable in this case.

The precipitation during April had been heavy, but it had also been heavy at many other times when there was no typhoid. During that summer all the river cities had typhoid, and in all cases the disease appeared at about the same time. At Port Huron, while there had been a tremendous excess of typhoid late in the fall of 1892, there were only one or two cases in the spring of that year; this pointed to a cause at or below Port Huron.

Finally, the speaker's attention was called to the Government dredging operations and to the place where the material was dumped. Mr. Williams. The flow of the Detroit River is given as 225 000 cu. ft. per second. From all available information as to the velocity of the water, it was found that it would reach Detroit in not less than 6 and would probably require not more than 10 days. From a careful study of the statistics of typhoid fever, as published by various Boards of Health, the conclusion was reached that the average period of incubation was 14 days. An examination of the statistics as to this disease showed that death, when it ensues, ordinarily occurs between the twenty-first and twenty-eighth days, which would give 25 days as an average. From April 16th, the day on which the first scow load of polluting material from the sewers of Port Huron was dumped into the St. Clair River, until June 5th, when there were four deaths from typhoid in Detroit, is 50 days. If 10 days are allowed for the water to flow, 14 days for the disease to incubate and 25 days to kill, the total is 49 days.

The investigation was carried on throughout the summer, and the curves of secondary and tertiary infection were studied, and finally, about seven weeks after cold weather set in and dredging operations were suspended, the typhoid ceased. Next year the typhoid appeared again after the dredging had commenced, and when it stopped the disease disappeared. These conditions have followed one another since that time, and investigations have shown that they prevailed in previous years. Going as far back as 1886, it was found that typhoid appeared during those years when dredging operations disturbed the channel of the St. Clair River or Lake St. Clair above the intake of the Detroit Water-Works, even as far away as Lake Huron; and the relationship could be traced accurately. In one or two years, when there has been no disturbance of this channel, Detroit has had a very low typhoid death rate.

Another interesting case occurred in 1897, which was a year of very low typhoid in Detroit, the rate being only 1.1 per 10 000 living. During that year there was only one group of cases which could be considered as epidemic, and that came in August. In two weeks there were seven or eight deaths. In looking for the cause it was found that the Government was dredging in the Clinton River. The work had been commenced some time before, and the dredge had been working gradually up the river, the material being dumped in Lake St. Clair, in what was supposed to be stagnant water, at a point about 16 miles from the water-works intake. At a certain date the dredge arrived at a bar in the river just below Mount Clemens, a town which, though not having a complete sewerage system, has a few sewers discharging into the Clinton River. A computation, similar to that previously described, showed that the group of cases in August was due to infection stirred up by these dredging operations.

Mr. Williams. These instances are given for the benefit of those who speak of streams which are not used as sources of water supply. There is no stream which is not a source of water supply for some person or community, and there is no danger of going too far in the restriction of the pollution of our streams.

Mr. Milne. PETER MILNE, Assoc. Am. Soc. C. E.—The American Water-Works Association has had the subject of stream pollution under consideration for the past eight years. Judging from correspondence with State Boards of Health and eminent medical authority, the speaker is led to believe that interstate legislation will not accomplish the object sought; and that it will be necessary to appeal to the Congress of the United States to seek the enactment of national laws to prohibit the pollution of water supplies. Such law may of necessity guard riparian rights and also protect the health of the people using the water of streams liable to be contaminated by sewage of every character.

A special standing committee of the American Water-Works Association has nearly ready for publication a voluminous report upon this subject.

An expression of opinion upon this important question by this Society will carry weight and influence and be accorded the consideration due.

An example of the manner in which a community may recede from its position taken regarding sanitation of water supply, is afforded by the experience of the City of Brooklyn, N. Y. More than four years ago, two streams, which formed a part of the daily supply, were condemned by the Board of Health of the City as unfit for use, on account of pollution due to the sewage of the population dwelling upon the water sheds, which pollution had increased yearly in proportion to the increase of population. These stream supplies were abandoned. In the present emergency and extremity of the city, viz., a long drought, it is suggested to again use the water of these streams, provided it is filtered.

The question of filtration being one of degree, it is apparent that its range of effectiveness in application is contingent upon the measure of pollution to be met and treated. If pollution is permitted to increase in intensity as well as quantity, it is reasonable to assume that the cost of filtration would be prohibitory and compel the consideration of new sources of water supply remote from the influence of pollution.

If the Federal Government enacts laws which in effect restrict pollution and contamination of the waters which are used by the people, it must invoke the skill of the engineer to aid it, on the one hand, and wise legislation, on the other hand, in order that equitable relations may be established and that the rights of all interests may be safeguarded in the effort to prevent the defilement of water supplies.

KENNETH ALLEN, M. Am. Soc. C. E.—In studying questions relating to stream pollution it is natural to turn to England, where, in official reports on the subject, there is a storehouse of information which is not to be found elsewhere; although of recent years the investigations undertaken by several States—notably Massachusetts—and by several cities, have made available much information of the greatest value. Increasing density of population, the concentration of large industries, and a general awakening of the masses to the importance of their sanitary environment, all serve to direct attention to the matter of stream contamination. Mr. Allen.

Up to the present time, the most important legislation on the subject has been the Rivers Pollution Act of 1876, resulting from a report of the Second Rivers Pollution Commission on the Rivers Mersey and Ribble, in which the following requirements, especially regarding organic pollution, which had been recommended by Dr. Frankland, were adopted:

"We suggest that the following liquids be deemed polluting and inadmissible into any stream:

"(*a*) Any liquid containing in suspension more than 3 parts by weight of dry mineral matter, or 1 part by weight of dry organic matter, in 100 000 parts by weight of the liquid.

"(*b*) Any liquid containing, in solution, more than 2 parts by weight of organic carbon, or 0.3 part by weight of organic nitrogen in 100 000 parts by weight.

* * * * * * *

"(*f*) Any liquid which, after acidification with sulphuric acid, contains, in 100 000 parts by weight, more than 1 part by weight of free chlorine.

"(*g*) Any liquid which contains, in 100 000 parts by weight, more than 1 part by weight of sulphur, in the condition either of sulphuretted hydrogen, or of a soluble sulphuret."

Although adequate provision for the enforcement of this Act appears not to have been made, its requirements were probably as fair and sound as was possible with the scientific knowledge of that time.

In the same year, 1876, the Massachusetts State Board of Health presented, in its annual report, a number of recommendations regarding the discharge of crude sewage into streams; among them:

"That no city or town shall be allowed to discharge sewage into any water-course or pond without first purifying it according to the best process at present known, and which consists in irrigation; provided that this regulation do not apply to the discharge from sewers already built, unless water supplies be thereby polluted; and provided also that such intended discharge can be shown to be at such point or points that no nuisance will arise from it."

But they are careful to add, that:

"The Board feels that, in the present state of our knowledge, sweeping laws for the general and immediate purification of all our streams would be hardly justifiable, and that they are not called for by the present condition of our rivers."

Mr. Allen. In the 23 years which have elapsed since these views were officially expressed, there has been opened up a vast field of research in sanitary matters, through the science of bacteriology, and, as our knowledge of the life history of these minute organisms is as yet quite imperfect, we are not likely to arrive at the ideal treatment of the matter for some years to come.

The standard of purity for sewage effluents, as established by the Second Rivers Pollution Commission, is faulty in the following particulars:

In the first place, the nuisance caused by a foul effluent depends quite as much upon the character of the stream into which it flows as on the effluent itself. A foul effluent discharged into a deep, swift-running stream of well-oxygenated water may cause no offence, while an effluent of half the size and twice the purity of the first, meandering over mud flats to a sluggish brook or mill pond may prove highly objectionable.

It is, therefore, seen that, if a standard of purity for the effluent is to be fixed, it should be in some way correlated to the character of the stream itself, and this under its most unfavorable conditions, as at extreme low water, when covered with ice, or when, for some other reason, it is particularly foul or devoid of absorbed air.

The two following examples serve to illustrate the inconsistency in the standard alluded to.

At the Leeds Congress of the Sanitary Institute of Great Britain, in 1897, Professor A. Bostock Hill, wishing to accentuate the difficulty of attempting to fix a standard for sewage effluents, quoted a case:

"Where a tank effluent, without any preliminary treatment, gave better results than the Rivers Pollution Commission standard, and yet subsequently so seriously polluted a stream that an injunction was applied for and obtained."

On the other hand, Alderman J. F. Spence (Chairman, Tyne Port Sanitary Authority), said:

"Between Tynemouth and Newcastle-on-Tyne, a distance of ten miles by the river, there is as large a number of works discharging foul and poisonous effluents as probably on any other stream in the kingdom. But these are so largely diluted by the enormous influx of sea water twice in the 24 hours, the difference in time between high water at the bar and Newcastle Quay being only 17 minutes, that the late Dr. Frank Buckland, when inspector of salmon fisheries, quoted the Tyne as one of the best salmon rivers in the kingdom, the pollutions being so diluted that the fish passed there unscathed to their spawning grounds in the higher reaches of the river. It may therefore be said the pollutions cause no nuisance, as the tide clears them all away."

Again, the Rivers Pollution Commission standard fails to appreciate the potential capacity for further purification in the effluent itself. This important point has been brought into prominence by the difficulty in securing permission, under existing laws, to discharge the effluents from certain bacterial treatment plants into neighboring streams, although it is generally admitted that no nuisance would result thereby. The outcome of this is that a Royal Commission has been appointed, and is now giving this matter careful study, which will no doubt end in placing before the world most valuable and complete information on this subject. Mr. Allen.

The final purification of sewage, which is due to the mineralization of its organic constituents, is effected by the action of minute organisms, either vegetable or animal, but generally by bacteria, and this is the case with crude sewage, the effluents from chemical precipitation works or from sewage farms, as well as from the septic tank or other of the methods known more specifically as bacteriolytic. To interrupt this action by the use of antiseptics simply postpones the inevitable and desired result, allowing organic matter to accumulate. Such interruption is, therefore, fundamentally wrong.

The process of mineralization may be roughly considered as taking place in two stages.

In the first stage, the solids are broken up and thrown into solution by liquefying bacteria, while the rearrangement of the molecules of all the organic matter renders it more susceptible to further change. This action is called bacteriolysis, and it takes place in a comparatively short space of time, the characteristic product being carbon-dioxide. Although usually occurring under anærobic conditions—when it is often accompanied by the evolution of sulphuretted hydrogen, marsh gas and other offensive or dangerous compounds, and sometimes, very likely, by the formation of those virulent poisons known as ptomaines—it is pointed out by Dr. Samuel Rideal as quite probable that it may result from the action of such ærobic liquefying bacteria as *B. Prodigiosus* and *B. Liquefaciens* under suitable conditions.

In the second stage, which is marked by oxidation of the nitrogen, and known as nitrification, the bacteria essential to the process must be supplied with oxygen, and it is characteristically an ærobic process.

It is true that in the formation first of nitrites and then of nitrates the oxygen required for the former may, in the absence of free oxygen, be wrenched from the nitrates or from other combinations of oxygen, leaving the final conversion into nitrates to be effected later.

Dr. Walter E. Adeney, of Dublin, has taken advantage of this action in devising the method of sewage and sludge purification known as the "Oxygen System," in which nitrification is effected by the application of nitrate of soda or permanganate of potash, substances which are capable of furnishing the oxygen required.

Nitrification is a comparatively slow process, and, while accomplished in an ideal manner by intermittent filtration, yet by reason of its slowness it may often be effectively carried on in streams by means of the oxygen dissolved therein. When, however, this element is not

Mr. Allen. absorbed from the atmosphere as rapidly as it is taken up by the organic matter undergoing purification, anærobic conditions prevail, which we call putrefaction, with their attendant evils.

It is clear, then, that the ability of the stream to oxidize the impurities in an effluent is a most important consideration, which has been quite ignored in the standard set by the Second Rivers Pollution Commission.

The determination of the organic nitrogen and carbon, as proposed by Frankland, has been found difficult in practice, and has been largely superseded by the less precise but more practicable determination of (or test by) albuminoid ammonia, as proposed by Wanklyn, or of the oxygen consumed, as proposed by Forschammer, or by both of these together.

Messrs. Pearmain and Moor have recently proposed the following standard, as adapted to present needs:

	Parts per 100 000.
Total suspended matter (not to exceed)	2.0
Albuminoid ammonia "	0.2
Oxygen absorbed "	0.714
Total hardness "	28.57
Nitrogen as nitrates (not to be less than)	0.714

They insist on the last requirement for the reason that, besides indicating the completion of the process, the nitrates also indicate a store of oxygen which may be drawn upon to oxidize the remaining impurities, insuring thereby a guarantee against putrefaction.

Dr. Rideal, to accomplish the same end, has suggested, as a safer test than any previously proposed, "the ratio of the oxidized nitrogen in the form of nitrate and nitrite to the 'oxygen consumed' figure," as the former represents the oxygen available and the latter the oxygen required to carry out the process of purification. He states as his belief that when the available oxygen is in excess,—

"Then under no subsequent conditions would a reversion of the sewage take place so as to cause a nuisance, and such a method of valuing the suitability of an effluent was independent of the size of the stream or the dissolved oxygen it contained. When, and in the majority of cases this was the case, the stream was many times the volume of the sewage and contained dissolved free oxygen, then this free oxygen was utilized instead of the oxygen of the oxidized nitrogen. Under these conditions it was not necessary for the oxidized nitrogen as compared to the 'oxygen consumed' figure to be so large."

If the character of a proposed effluent, as well as its probable dilution in running water, could be predicted with certainty, its liability to cause offense could be pretty well foretold. A rough but convenient limit was determined in this way by F. P. Stearns, M. Am. Soc. C. E., when Chief Engineer of the Massachusetts State Board of Health, substantially agreeing with a similar limit fixed by Rudolph

Hering, M. Am. Soc. C. E., with reference to Chicago's sewage discharge, in which he designates a stream flow of not less than 2½ cu. ft. per thousand population as sure to cause offence, and a flow of over 7 cu. ft. per thousand population as not likely to cause offence. Mr. Allen.

Such a rule will, of course, apply only when manufacturing wastes are negligible in quantity or character; but where questions involving the dilution of crude sewage are concerned, this method of predicating offensive conditions is likely to prove as convenient and satisfactory as any, considering the number of indeterminate factors involved.

Whether more definite tests, to cover various conditions, are practicable remains to be seen, but the more recent ones appear to be steps in the right direction. Possibly the future may discover some simple biological tests by noting the effect of polluted streams upon various kinds of fish or other organic life contained in them. In fact, in the elaborate series of experiments on sewage disposal which has been carried on for several years past at Leeds, one test of certain effluents was their capacity to support the life of carp, the effect on confervoid growths being also noted. It is known that different kinds of fish, such as salmon, bass, catfish and menhaden, require waters of different degrees of purity, and their survival would, to a certain extent, be a measure of the contamination of the water in which they were placed.

A great economic advantage in bacterial treatment is the elimination of the vexed question of sludge disposal. It is true that sand and other grit still have to be removed, and that woody fibre requires time to become peptonized without clogging the filter, but the troublesome and offensive organic solids may be effectually disposed of, and this is a great step in advance.

The relative efficiency of such treatment, as compared with mechanical filtration and chemical precipitation, is well illustrated by the following results obtained at Leeds:

	Oxygen Absorbed in 4 Hours.		Albuminoid Ammonia.	
	Parts per 100 000.	Percentage purification.	Parts per 100 000.	Percentage purification.
Crude sewage	12.14	....	1.21	....
Filtration through paper	5.53	54	0.67	45
Lime precipitation	6.57	46	0.64	47
Bacterial action (Dibdin's filters)	0.91	92	0.125	90

Crude sewage contains several million bacteria per cubic centimeter —London sewage some 3½ million. After complete nitrification, these are reduced to an insignificant number; but before that time, although an effluent may contain sufficient oxygen to complete the process of purification, it may contain a very large number of these organisms.

Mr. Allen. Now, while there is some evidence to indicate that pathogenic varieties will have been eliminated before arriving at this stage, yet with the present imperfect knowledge, and with the possibility of their subsequent introduction into an effluent yet containing nutrient material, it would probably be wise to fix a low limit to the bacteria per cubic centimeter permissible in a stream used for a water supply which has received the effluent from biological filters. In fact, it is a question whether the water of such a stream should not be filtered before distribution to a city, as there is always a chance that at times, through accident or excessive flow, the purification of the sewage may not be up to the established standard.

When the stream is not used for drinking purposes, the presence of a large number of bacteria in the effluent is not necessarily an unfavorable indication.

It is believed that, for the present, legislation regarding stream contamination and sewage purification should be largely tentative, both because experience in these matters is as yet limited, and because it is rapidly increasing.

In dealing with a subject so complex and, at the same time, often involving the public health and the outlay of large sums of money—in the establishment of purification works on the one hand and the depreciation of riparian property on the other—each case of pollution should be considered by itself, by an authorized official, who should be a man of much tact and ability, as well as one having special knowledge and experience in his line. He should be guided by certain general laws regarding stream pollution, but should be given considerable discretionary power, and, to avoid troublesome changes of policy, it would probably be well to have him serve a long term of years. On the other hand, he should not be enabled to use his authority in too summary or arbitrary a way, to the disturbance of vested interests, when not absolutely necessary for the public good.

In order to make his supervision effective, the area under his control should be defined by the natural water-shed lines; that is, he should have authority over the entire drainage area of one or more streams, and be responsible for their condition.

In order to effectively control stream contamination, it is suggested that all streams may be placed in one of three general classes, viz.:

I. Streams far from populous centers and not likely to be used for manufacturing or domestic purposes.

II. Streams likely to be used for other than domestic purposes.

III. Streams likely to be used for domestic purposes.

Such a classification would not appear to be difficult or intricate, and laws regarding permissible pollution would apply, in general, to all streams of the class in question. Class I would require little or no legislation, while certain standards of purity for sewage effluents,

Mr. Allen.

etc., would apply to Class II and others to Class III. Moreover, in the case of large rivers, like the Mississippi or Hudson, it would be quite possible to include the headwaters and upper reaches in Class III and leave the lower reaches in Class II. On the other hand, if certain populous centers toward the mouth of the stream should befoul its waters to the detriment of the fishing or other interests near by, it might be necessary to create a special district to which the restrictions for Class III would apply.

That the purification of the sewage of large cities by land disposal is feasible, is seen by such examples as Berlin and Paris.

Suitable land, however, is not always available, when chemical treatment, followed by filtration through the natural soil, as at Birmingham and Manchester, or through prepared filter beds, as at Glasgow and Sheffield, may be resorted to. In the cases of London and Salford, the sewage is merely clarified by chemical precipitation, and the sludge is carried out to sea by steamers and dumped; and in such cases the effluent cannot be said to be truly purified. The previous examples, however, suffice to show that there is no inherent difficulty in purifying the sewage of a large city, even if irrigation cannot be availed of. It is merely a question of selecting the best and cheapest.

The modern bacteriolytic processes avoid the acquirement of extended areas for irrigation on the one hand and the question of sludge disposal on the other, and from them the ideal solution of the sewage disposal question of the future may be hoped to be realized. Experience has already been sufficient to obtain for these methods very high endorsement.

For instance, in testimony before a Local Government Board Inquiry at Exeter, regarding the septic tank treatment, Dr. Rideal gave as his opinion, that

"Wherever a sewage farm would work, this system would work, and anything which would prevent this system working would prevent a sewage farm working."

When asked whether it could be applied to very much larger towns, he replied "much larger." Dr. Dupré, also, declared that "no system of purification hitherto known had such a complete effect upon sewage."

It must be admitted, however, that up to the present time no great city uses any of these methods. The largest plant is at Sutton, England, where some 145 000 U. S. galls. per day are treated by the Dibdin system of double filtration.

The Sewerage Commission of the City of Baltimore, which for some time has had this very matter under consideration, while admitting the admirable results attained at Sutton, Exeter, and elsewhere, and after consultation with Mr. James Mansergh, regarding this particular point, has refrained from recommending any one of these sys-

Mr. Allen. tems for so large a city, involving, as it must, much that would be new when applied on the scale proposed. But this decision is qualified by the statement that

"The Commission fully recognizes that one or more of them may prove available for the treatment of the sewage of some of our outlying districts, and may, if future experience sustains the claims and anticipations of their advocates, yet become adequate to our service on a larger scale."

To sum up what has been said, it is believed:

(*a*) That stream contamination should be prohibited, and the purification of the sewage of cities be compulsory only:

(1) When it is offensively evident to the senses of smell or sight.

(2) When manufacturing, fishing, or other productive interests are likely to be seriously affected thereby.

(3) When the health of neighboring communities is jeopardized, either through the authorized use of the stream for domestic purposes, the probable contamination of areas devoted to the culture of shell-fish, or the probable formation of deposits upon shoals or shores.

(*b*) That the purification of the sewage of large cities is always feasible. It is always costly, but it is probable that the introduction of bacteriolytic methods of treatment will, in general, when more fully developed, tend to economy.

(*c*) That to make legislation effective:

(1) All streams should be classified according to their permissible contamination.

(2) A responsible officer should supervise and control all sources of pollution throughout the drainage area of every stream in which pollution is restricted by law.

(3) Where possible to formulate consistent regulations regarding the limits of contamination, this should be done; but that with regard to organic contamination, in the present imperfect but advancing state of our knowledge on this subject, the matter may best be left largely to the discretion of the official in charge of the particular drainage area in question.

Mr. Owen. JAMES OWEN, M. Am. Soc. C. E.—It has not been regarded as within the province of this Society to discuss legal questions, but, recognizing legal enactments as the crystallization of public opinion on the subject legislated upon, the legal aspect of stream pollution is worthy of consideration. The radical, unqualified proposition that all pollution of streams should be summarily stopped seems to the speaker unwise. The question is a very live one in New Jersey, and the conditions in the northern part of the State furnish an object lesson.

The Passaic River was formerly known as "the beautiful Passaic;" now it is called "the dirty Passaic." Along its banks is a population of about 400 000 persons, who, by the construction of sewers, have completed, within the past twenty years, a system of discharge which renders the river unfit for potable uses, or for any purposes for which water can be used. Until three years ago, Jersey City took its water from the lower Passaic. It was bad water, much contaminated and unfit to drink, and Jersey City vainly tried, for years, to obtain from the Courts means to restrain the towns above its intake from discharging their sewage into the stream. Mr. Owen.

Newark obtained a new water supply about ten years ago, and Jersey City got a new supply three years ago. Curiously enough, as soon as Jersey City stopped taking its drinking water from the river, there arose a popular demand that the pollution of the river should cease. Paterson, with 100 000 population, is discharging a large amount of sewage into the river, and Newark and the towns between Newark and Paterson wish defilement from this source prohibited. Newark also discharges an enormous amount of sewage into the river, and this drifts up with the tide for four or five miles above the city.

On the Upper Passaic is the small town of Summit, which is discharging into the river, not crude, but filtered, sewage. The City of Paterson has brought suit against Summit to prevent its pollution of the stream. As a kind of cross-fire comes the proposition of the City of Paterson to enjoin the East Jersey Water Company from taking water from the river, on the plea that there will not be left in the river sufficient water to take the city's sewage down. From this it will be seen that the trend of public opinion in that locality is indefinite.

It cannot be too strongly emphasized that there should be no contamination of drinking waters or the sources of such waters; but the matter of the discharge of sewers into rivers used merely for navigation, or for drainage, should be considered as a question of dilution; and if, within the next few years, a standard of permissible dilution, which would be free from effects deleterious to the community, can be established, the prime object of this inquiry will have been attained.

Charles G. Darrach, M. Am. Soc. C. E.—The study of this problem for twenty-five years leads the speaker to agree with the conclusions so admirably expressed by Mr. McMath. The engineers of America are competent to solve this problem, and the speaker anticipates not only its solution but such a development in the future as will modify the water-supply question to the extent that if necessary a city will not be obliged to go beyond the purification of its own sewage to supply itself with water. Mr. Darrach.

E. W. Harrison, M. Am. Soc. C. E.—The subject of the right of cities on tidal streams to discharge their sewage therein has not been taken up in this discussion, but some recent events indicate a disposi- Mr. Harrison.

Mr. Harrison. tion to impose unnecessary and harmful restrictions on such communities. When the City of Newark was founded, about 200 years ago, the site was chosen because it was on a tidal stream. It offered, by reason of its location, advantages for settlement, for manufactures and for commerce. This agitation for the purification of the Passaic River, of which Mr. Owen spoke, resulted in swinging the pendulum far from the stand taken by the Courts ten years ago, for in March last the City of Newark was enjoined from draining about 130 acres of land into the Passaic at a point where the tide ebbed and flowed. The water at this point was salt, though not as salt as the sea, but both medical testimony and a bacteriological examination were submitted to show that the growth of bacteria was inhibited by reason of its saltness. The river was thus acting only as a carrier, conveying sewage to the sea without affording opportunity for the further growth of bacteria.

That agitation and decision are now bearing strange fruit. Other towns draining into estuaries of New York Bay are seeking places of deposit for their sewage, and they can keep their water supplies pure only by turning their sewage into the tidal streams. This decision has helped to create the opinion in northern New Jersey that no sewage should be emptied into any stream, tidal or otherwise, not even into the Kill von Kull, a tidal strait between Staten Island and New Jersey, passing an enormous quantity of water with a swift current.

The recognition of this argument as a governing principle would operate to deprive cities located on estuaries and tidal streams of one of their great advantages as manufacturing sites. If a manufacturing plant cannot be established along the Arthur Kills, or a large city grow up there with the right to discharge sewage into the salt water; if investors in such enterprises must contemplate an enormous bond issue to take care of sewage, the manufacturing plant or the city might as well be located up in the mountains. Engineers will discriminate between drainage into salt water and drainage into a fresh-water stream, and, while the speaker agrees with Mr. Owen that the pollution of streams now used or likely to be used for water supply should be prohibited, the same arguments do not apply in the case of tidal streams; otherwise sewage purification works would be required for cities like New York and Brooklyn.

Mr. Maignen. P. A. Maignen, Esq.—In Paris the distribution of the water supply is double and of two qualities; raw river water for industrial purposes, street sprinkling, fire extinguishing and other base uses—sold at 12 cents per 1 000 galls.; and spring or assimilated (filtered) water for drinking and household purposes—sold at 24 cents per 1 000 galls.

The potable water distribution, being kept at a higher pressure than the industrial water, became largely used for hydraulic elevators and other purposes as motive power. In order to discourage this

Mr. Maignen.

misuse of the pure water, the rate for such purposes was raised to 48 cents per 1 000 galls. This differential rate has prevented the abuse of the pure water, so that filtration plants have to be provided only for water legitimately used and well paid for.

Another reason for which French engineers advocate the double canalization system is to maintain an even pressure in the pure water distribution, so as to reach the highest points of the cities at all times. It is customary, in many towns, to water the streets and flush the sewers, for several hours every day. With a single canalization the upper parts of the city would be deprived of water by the simultaneous opening of the low-town hydrants; this trouble does not occur with a double canalization.

Filtration, as an art, is still in its infancy. It is a mistake to ask too much of a single filtering operation. The work should be divided. Something has already been accomplished in this direction by the use of storage reservoirs, but unless they are, as in London, large enough to render pumping during storms unnecessary, they are only extensions of the river, and of little use. If storage reservoirs are wanted in case of accidents to the pumps, by all means build them, but, in the speaker's opinion, clarification can be obtained equally well by quick upward filtration through coarse sand or other coarse material—the choice varying according to the particular quality of the water to be dealt with—and at less expense.

Sand filtration, as carried on in Europe, with sundry improvements, could be applied as a secondary operation. And it is possible to find, in the large industries, such as sugar and chemical works, breweries, wine producers and the like, which have recourse to filtration, some ideas which may be of service in carrying out a third and final operation in the purification of water for city supplies.

Mr. Tribus.

L. L. Tribus, M. Am. Soc. C. E. (by letter).—For twenty years past the providing of suitable water supplies for thickly settled districts has been receiving more and more intelligent consideration; the relation of physical health to the wealth and efficiency of a community has been more carefully studied from the economic as well as the æsthetic standpoint; and to the ends at issue have been brought the brains, energies and conscientious efforts of engineers, chemists and even of municipal officials themselves, who have been popularly classed as men devoid of other than political instincts.

To a more limited degree, the disposal of the wastes of civilized life, passing off as sewage, has received careful attention, and is beginning to be treated as it deserves, as one of the most important subjects for the engineer to consider. It is not alone necessary to provide human beings with pure food and drink; it is equally necessary to remove the refuse after the valuable parts are used.

The questions presented for these informal discussions, relating to

Mr. Tribus. Stream Contamination by Sewage, Purification of Sewage, and Filtration of Water, are so correlated as to make their separation here scarcely wise, although certain features of each are distinct.

By far the larger number of cities (the term "cities" including all communities making use of municipal systems of water supply or sewerage), take their supply from rivers or lakes, larger or smaller as the case may be; those using the various forms of driven and artesian wells not, as a rule, having the classes of waters that need artificial purification. This being the case, it is evident that the source of supply should be maintained in all possible purity, rather than that the water should be permitted to deteriorate, and then need to be purified by some form of filtration, and at great expense.

Consequently, defiling matter, as certainly raw sewage must be considered, should not be admitted to a lake or stream if there is any possibility of injuring the water supply of another place. There is no justice in permitting one community to save itself trouble and expense by entailing trouble and expense on another, or in event of that other not having sufficient enterprise to protect itself, and perhaps furnishing to it elements of disease and danger. Rights to the waters of a stream have caused litigation for centuries, but it is only in comparatively recent times that "use" of the stream has been considered as requiring, morally, if not always legally, passing it on in unimpaired purity.

Certain classes of sicknesses known as "Filth" diseases, of which typhoid is the most prominent, are usually attributed, and probably with justice, to the dissemination by the water supply (or milk, directly associated, too often, with it) of germs contributed by sewage to the sources of the water system. Dilution of sewage with water has been largely looked upon, even by engineers, as a means of purification, and to some extent it is, but dilution does not destroy germs; and disease germs in a large volume of water have just so much more chance of dodging the other forms of bacterial life which seek to destroy them.

Not many years ago, purification of sewage was not an unmixed success, chiefly because of unintelligent design and carelessness in the operation of plants, the method of Nature's action not being entirely understood, or, if so, then ignored; the germ theory had scarcely been promulgated, and the existence of bacteria of different kinds, which, like people, preyed on each other, was unknown.

It would be a difficult matter to estimate in dollars and cents the losses of property, the destruction of valuable material, and the injury to lives, directly traceable to the emptying of crude sewage into streams; yet few will deny that the aggregate sum would be vast. The driving away or destruction of fish life, removing thereby a very valuable article of food; the depreciated value of property adjoining such sewage contaminated streams, and the actual deaths of hundreds or thousands of human beings, from the use of polluted waters, sufficiently indicate the conditions at stake. The latter topic has been so admirably covered in recent papers on typhoid and allied diseases, with their relationship to water supplies, that merely speaking of the remarkable reduction in death rate, noticeable after the introduction of pure supplies, is all that is necessary here. **Mr. Tribus.**

The New York State Board of Health, in its policy of saving streams and lakes from further contamination, practically compelling cities to treat their sewage, seems to the writer to have taken the only proper action that justice and right can countenance; yet there are instances where such exclusion would appear an unnecessary precaution, for the rights of no one would seem to be injured by "direct" disposal.

It is a question, therefore, that had best be left to the intelligent and more enlightened sentiment and judgment of communities directly concerned, trusting to fully sufficient existing laws to prevent undue trespass on the rights of others. The writer believes, however, that the proportion of sewerage systems to be constructed in the future, without adequate disposal and purification plants, will be very small; and that a large number of the systems now in use will ultimately require treatment.

The same discussion applies very largely to the second question—"Should the Purification of the Sewage of Cities be Compulsory?" Present laws or court decisions cover fully all cases where non-treated sewage becomes a nuisance to residents of the city itself or to those of other places. Unfortunately, enforcement of laws seems to be a very difficult matter, and there is no reason to suppose that any new laws, even apparently more binding, would work to any greater advantage.

As far as applicability to large cities is concerned, new questions always arise. In almost all instances, large cities have grown up on the shores of tidal harbors, great lakes or large rivers, so that sewage flow bears usually so small a proportion to the volume of moving water, fresh or salt, that the dilution dissipates most of the cause of offence, though in the course of years, unmistakable and permanent indications of defilement can be noticed near the outlets of the sewers.

It is necessary to note clearly, what purification of sewage actually means. It is the total resolution into harmless inorganic elements, of all defiling or organic matters—a complete chemical combustion or oxidation. In this process, bacteria, abounding in the sewage itself, have a great deal to do; in the opinions of some, perhaps more than any other agent.

A direct answer as to its feasibility for large cities is impossible; each city must be judged by its own conditions. New York, Philadelphia and Chicago would illustrate the impossible, or at least, impracticable. Paris has, in a large measure, successfully embraced purification. London aims to make the effluent inoffensive, but it is

Mr. Tribus. far from being purified; its chemical treatment and subsidence being only an arresting of Nature's processes, and not a completion of them. To a thorough understanding of this last subject must those come who have the handling of sewage problems. Nature cannot be hurried, though she enjoys being assisted. We must conform to her laws or failure will ensue.

It is possible to purify sewage, and it is practicable, but it takes land in large areas to do it economically, and land properly selected and prepared, with the utmost care in its use, besides; or it takes water in such large volume that the same processes of nitrification and oxidation can be carried on without regard to time, and be conveyed away by currents or streams, so as to be inoffensive during the operations. The methods of treatment need not enter into this discussion; but it must be borne in mind that, unless local conditions absolutely prevent, some form of land treatment, without the use of chemicals, comes nearest to Nature's own most efficient process, and consequently is most likely to prove successful, even under severe climatic conditions.

Closely allied to the purification of sewage, and even more important to human life, is the providing of pure water supplies. Mountain lakes and brooks in uninhabited or sparsely settled districts can usually be taken as of sufficient purity, if the water be intelligently gathered and cared for.

The trend, however, of growth of population, is in the grouping of communities along the courses of the larger streams and on the shores of lakes. This entails more care in the selection of intake points, and is gradually tending to make undesirable the use of neighboring waters for public supplies without treatment before use. On the other hand, the smaller streams and lakes, not subject to pollution, do not furnish sufficient quantities of water for the larger cities, even when storm-waters, in addition to normal flow, are carefully conserved. It is not always feasible to secure ground-water or rock-well supplies, nor are such waters always safe or desirable themselves, so that recourse must be had in most cases to surface sources. Most persons will grant that a surface supply, from unimpeachable collecting grounds and stored in well-kept, clean reservoirs, is all that can well be desired, but in the more closely settled States, each year, such a condition is becoming much harder to find. Then, again, in the use of surface waters in their natural state there is always somewhat of an uncertainty, for, in spite of the most careful supervision of the water-shed, unless it be absolutely uninhabited, contaminating elements are always likely to appear.

The writer is not quite ready to insist that all public supplies be filtered, for here, as in sewerage questions, situation has a great deal to do with the method proper for adoption; but certainly, intelligent filtration never harms a water, while its lack may entail danger. Filtration, in some form or other, undoubtedly began with man's use of water; its adoption, on strictly scientific lines for public use, is a development of very recent years. We may neglect consideration of the small household filters, the thousand and one devices that improve or injure the water according to the care in their use, and take up the broader field of filtration of municipal supplies as a whole, by the two general systems which may be classed as "Chemical" and "Natural." Mr. Tribus.

It is unnecessary here to discuss the comparative merits of either. It is sufficient to say that both have full fields for operation, the main point being that it is fully proven that impure waters can be filtered and rendered safe and desirable for domestic use. At the present day practically everything is measured by its value in dollars and cents, a materialistic theory which we have to recognize even while combating it as a principle, so that the *pro* and *con* of filtration for any particular city seems to be: Does the advantage of a purer water, in decreased death rate of population, figuring life at a minimum value, or a gain to the community of the work of so many persons annually, equal the operation, interest on cost and appropriations for a sinking fund for a filtration plant? Of course, taxes and politics, and various other, often conflicting, interests are also considered; though impaired health, more or less discomfort, and offensiveness to eye and nose, seem to have but little weight. Usually the activity of a violently deadly epidemic is necessary to impress the weight of the argument.

The writer is amply convinced that the coming solution of the public water question, considered as a whole, is in "purification," implying rather more than the narrower term "filtration," though that, in its growing ordinary acceptation, includes the very complex operations of Nature which result in thorough purification, hence safety for the health of, and satisfaction in other uses by, the public at large.

Palmer C. Ricketts, M. Am. Soc. C. E.—(by letter).—The great value and advantage of maintaining, if it were possible, all streams, large and small, in the purity they exhibited before there were settlements upon their banks is self-evident. The fact, however, remains that, to a greater or less extent, streams always have been and always will be natural sewers. They will be as long as water keeps flowing from the land into them and carrying with it impurities dangerous to health. Those flowing in thickly populated valleys certainly will be, even if the ordinary sewage of the cities on their banks be purified before being discharged into them. The fæcal matter of one inmate of one farm house may cause more havoc than a year's sewage flow from a large town. Mr. Ricketts.

As an example, consider the Mohawk River Valley in New York; narrow, with a long line of villages, towns and cities along its banks, and elsewhere quite thickly settled. If no sewage from towns and cities were permitted to be discharged into it the water would, never-

Mr. Ricketts. theless, be unsafe for drinking purposes, if not filtered. This river discharges into the Hudson River below the intake of the Troy waterworks. Below Troy, the cities of Albany, Hudson and Poughkeepsie filter the water pumped from the Hudson River. At least the last two have filtered it for years and the first will shortly have its extensive plant in operation. Under these circumstances it would be extremely difficult to force the great population of the towns and cities in the Mohawk Valley to purify their sewage before discharging it into the river, in order that its water might be more pure, but not pure enough to be drunk with safety.

In the "Report on the Chicago Drainage Canal by the Commission appointed by the Mayor of St. Louis, Mo., 1899," may be found the paragraph:

"By general law of the United States all cities must, sooner or later, be prevented from polluting the natural water courses of the country with sewage."

However much jurisdiction the United States may have in the Chicago Drainage Canal question, it would have to be considerably stretched to cover the case of a non-navigable stream, like the Mohawk, lying wholly within a single State and discharging into a river lying, as far as that part of it valuable for water-supply purposes is concerned, in the same State. And being thus practically a State question, how much chance is there that this large part of the population can, by act of Legislature, be put to great expense without proof that it is necessary.

When the writer became engineer of the Public Improvements Commission of the City of Troy, N. Y., and had, in consequence, to design a complete system of sewerage for the city, he was asked whether it would not be advisable to build an intercepting sewer along the river and purify the sewage of the whole city at the south end before discharging it into the Hudson River. This would have entailed pumping 9 000 000 galls. a day, exclusive of any rainfall, to a height of about 200 ft., purifying it, and discharging the effluent into the Hudson about 5 miles below the mouth of the sewage-laden Mohawk River. Twenty-five years ago the City of Troy began to pump most of its water supply from the Hudson above the mouth of the Mohawk, where the water was quite pure. Instead of doing this it could have obtained a bountiful supply by a gravity system at a reasonable cost. In the meantime the river has become more contaminated above the intake, by the sewage of towns which have since grown larger, and the quite pure water supply still exists in the highlands back of the city. Under these circumstances could Troy, in equity, compel the towns above to purify their sewage?

The writer has thus taken up the conditions existing in one neighborhood to show that no radical general rule can reasonably be made regarding sewage purification. General laws, absolutely prohibiting the discharge of unpurified sewage into streams, would in many cases be destructive and unjust; and, in the design of a system of sewerage, it is well to remember that it is easy to put a community to unnecessary expense to meet possibilities which never materialize, and sometimes to prevent the construction of much-needed works by prohibitive and unnecessary conditions. Mr. Ricketts.

On the other hand, it is not necessary to state that many cases exist where the purification of sewage before its discharge into streams is advisable and necessary. This, of course, is particularly the case where the streams are small and the country thickly settled. Those in the neighborhood of Boston afford a good example. It also seems evident that even if a stream is not to be used for a water supply it should not be permitted to become a nuisance because of sewage discharged into it.

As to the pure-water question for cities, the writer believes filtration to be its coming solution. More or less pure water may be used without filtering if it comes from a source uncontaminated by the sewage of villages or cities, but the only way to be sure that it shall be reasonably pure is to filter it.

James Smith Haring, M. Am. Soc. C. E. (by letter).—There is a phase of the discharge of raw sewage into rivers and streams which was not mentioned in the oral discussion of this subject, that of the minimum flow, in either rivers or streams. Mr. Haring.

This is a matter of great importance in considering the question of permitting untreated sewage to enter sources of water supply, not only as regards the effect upon the purity of the supply for other localities, but also the relation which the amount of sewage passing into the stream has to the quantity of water flowing therein, to prevent the nuisance of the collection of undiluted sewage in the immediate vicinity.

Most western rivers are subject to large variations in both depth of flow and area covered during the periods of maximum and minimum seasons. Although the volume of water is calculated from the total quantity passing in the channel, yet it may be so divided and diverted as to leave the conditions attending the dilution of sewage quite at variance with the permissible allowance. Especially is this true of rivers having very rocky channels, in which the flow is concentrated at certain points into small cross-sections; while, at other places, the water is spread out over a very shallow but wide and stony bed and almost disappears. This is also true of a river of sandbars and marshes. The collection of sewage in beds of this character certainly emphasizes the necessity of compelling its treatment before discharging it into streams used for water supplies. It seems quite unnecessary to require any treatment of sewage before disposing of it in tidal

Mr. Haring. salt-water bodies. Cases might arise where sewage might be caught on a flood tide and carried back upon the shores and cause a nuisance, but the conditions would not be general, and the writer does not think that in the discussion of this subject such a contingency was anticipated.

The only way to make the purification of water supplies effective is to first prevent, by Federal laws, the pollution of rivers and streams. Too many of our rivers are in the jurisdiction of more than one State to make the State control of these matters effective, and the personal feelings of lawmakers are too largely acted upon by the fear of giving offense to some person or interest in their constituency to make a success of individual State legislation. Then, too, individual cities and towns will procrastinate to a degree almost, if not quite, criminal, rather than press on to a reform, or even to provide the means necessary to remedy or prevent the pollution of streams, because of the attending expense; and though conscious and confessing the, shall we call it, crime, permit it to go on. Only a general law, irrespective of State or city, can prevent effectually what will be in a few years, if it is not in some places already, an evil of great magnitude.

Filtration is destined to be the important agent to prevent wholesale contagion from water-borne diseases, even though it be compulsory by law to treat sewage before permitting it to flow into streams, and even though the effluent be purified "to a drinking-water standard," as one business concern engaged in that work now claims to do (though the standard may need qualification). The rapid development in the education of the people demands the filter as a guaranty of the best that can be had in water supply. Though sewage may be treated and made inoffensive by the expenditure of money, yet water supplies cannot be manufactured, and the daily increasing news of water famine all over the country admonishes us of such conditions.

Mr. Fuertes. JAMES H. FUERTES, M. Am. Soc. C. E. (by letter).—A brief analysis of what is already published on these subjects will be, it seems to the writer, of more value than the addition of new compilation to obscure further the simple principles governing the points under discussion.

The right of polluting streams, under certain conditions and to a certain extent, has been a large factor in the progress of all advanced countries. In some of our States the Mill Acts, passed to foster growing industries, would be inoperative if this right were not conceded. The term pollution is always comparative. No water found in Nature is pure, the purest being that from mountain springs. Going down the scale in the order of comparative pollution, would then be classed ordinary ground-waters, waters impounded in large storage reservoirs, waters of large normal rivers, waters of large lakes taken some distance from the shores, waters of upland streams, and, lastly, wells and bodies of water polluted directly by sewage matter near the point of abstraction.

Mr. Fuertes.

Prevention of pollution must, therefore, imply degree of prevention desirable, because all streams are already polluted by natural agencies beyond the control of man's legislation, unless flies and birds and winds and migratory animals can be legislated out of existence.

To prevent stream pollution absolutely would be undesirable. If a stream is not to be used for a water supply, or if its waters are not to enter into the composition of matters intended for human consumption, sufficient pollution prevention to insure its not being offensive in the slightest degree, or injurious to manufacturing interests, would be sufficient. This can be secured under the existing laws. If the water is to be used for a water supply, its protection must be more complete; but here, again, the laws are ample in scope. Not only would it be undesirable, but it would be impossible to secure the passage of absolute pollution-prevention legislation, because such laws would paralyze the manufacturing interests of the whole country, and would thereby arouse irresistible organized opposition. These views are expressed by the best expert opinion and by legislative bodies.

The discharge of a large amount of sewage into a small stream will, undoubtedly, make it offensive and unfit to drink; but the discharge of a much greater amount into the Upper Mississippi River, would be undetectable at its mouth. Thus, at New Orleans, the effect of the sewage discharged into it and its tributaries by all the cities between the Alleghenies and the Rockies is completely masked by the much greater relative pollution of the same kind, caused by surface washings from the water-sheds of these streams. In this connection, it will be interesting to observe whether or not the influence of the Chicago sewage, after it is directed into the Mississippi River, can be detected at St. Louis.

As a health protective measure, for cities on large rivers like the Mississippi, Ohio, Hudson, etc., it would be more efficacious and less expensive for all to discharge their crude sewage into the rivers and then purify their drinking water, than for all to purify their sewage and not their water supplies. On smaller streams both processes may be necessary. So far as the prevention of stream pollution is concerned, the laws are adequate; witness the recent cases of Danbury, Conn.; New Britain, Conn.; Santa Rosa, Cal., and numerous others.

The writer, in a statistical study compiled in 1896,* called attention to the extent and importance of pollution from surface washings, and since then has had verification of his deductions by seeing the same views expressed in the findings of a special investigation for the War

* "Water and Public Health. The Relative Purity of Waters from Different Sources."

Mr. Fuertes. Department on the cause of typhoid fever in the army camps during the late war, and in the report of the board of experts appointed to report on the improvement of the water supply of the City of Philadelphia.

The indications are that, with a little more popular education in these matters, the people will, before many years, demand the purification of all waters obtained from surface gathering grounds, rivers or lakes.

The purification of the sewage of large cities is not always practicable. So far as safe-guarding public health is concerned, it is vastly more important to sterilize all the air we breathe than to purify the sewage—for instance, of New York City. One would be just about as feasible, from a financial point of view, as the other. Probably there are, at least, one hundred times more deaths in the city resulting from breathing impure air than can possibly be traced to the combined effect of the sewage of the city and the infection of water and food supplies.

Filtration, if properly planned and conducted, is the best and most satisfactory way to purify drinking water, so far as we know, to-day. What the next fifty years will bring forth, however, is beyond the power of any mortal to prophesy. In the past two decades, so many of our standards have been changed, and so many useful scientific inventions have been perfected, that great advances are to be expected in every field.

There may be individuals in the next generation who will live to see 10 000 000 people resident in Greater New York. When that time arrives the water needed by the population will equal the present low-water flow of the Hudson River at Albany—draining over 8 000 square miles of country. The writer does not believe that any one now living can say how that water will be derived.

photo courtesy Williamsiana Collection,
Williams College

Robert Emmet McMath

Robert Emmet McMath was born in Varick, New York, on April 28, 1833. He attended Williams College (as a fellow student of President James A. Garfield), graduating in 1857. In 1858 he began his engineering career in St. Louis with the design and reconstruction of a levee at the confluence of the Missouri and Mississippi Rivers. From this beginning he remained in active practice for most of the remaining sixty years until his death on May 31, 1918. His long and varied career continued as Deputy County Surveyor of St. Louis County, Missouri from 1859 to 1862. This was followed by employment as an Assistant Engineer with the United States Coast Survey. During this time McMath made a topographic survey of what is now Arlington National Cemetery, and a survey for the Nicaragua Canal which included about 1000 miles of the San Jaun River, then bordered on both sides by unpenetrable wilderness. In 1865 he became an Assistant Engineer for the U.S. Engineer Corps. In this position he conducted surveys of the Illinois and Arkansas Rivers before being placed in local charge of Illinois River improvements as part of a canal and slack-water navigation from Chicago to the Mississippi River. While so engaged McMath pioneered in the use of brush revetment for large rivers. In 1872 he moved up to the Mississippi River and in 1873 was appointed Principal Civil Assistant. He remained in charge of river improvements and various special physical investigations culminating in a position with the Mississippi River Commission from 1880 to 1883. To quote his contemporaries "During the period of his association with the Mississippi River Commission, Mr. McMath was regarded as the best informed engineer on river hydraulics in the country."

With no previous experience in sewer work, Robert Emmet McMath was appointed Sewer Commissioner of St. Louis in 1883, a position he held until 1889 when he "was retired to private life by an administration of opposite politics". During this period he was involved with many aspects of sewer work. One of the many papers he presented before ASCE was "Determination of the Size of Sewers". The formula presented in this paper to estimate storm discharge from drainage basins exceeding 100 acres was widely accepted by engineers and became known as the "McMath formula". After a period of private practice he was elected President of the Board of Public Improvements of St. Louis, a position which he filled from 1893 to 1901. The remainder of his active career was spent in private practice.

He was a prominent member of the Engineers Club of St. Louis, serving as President in 1886. He served as a Director of ASCE in 1889.

AMERICAN SOCIETY OF CIVIL ENGINEERS.
INSTITUTED 1852.

TRANSACTIONS.

Paper No. 945.

THE SANITARY DISPOSAL OF MUNICIPAL REFUSE.

AN INFORMAL DISCUSSION.*

By Messrs. Rudolph Hering, John McGaw Woodbury, H. de B. Parsons, W. F. Morse, George A. Soper, Theodore Horton, W. J. Baldwin, Charles B. Ball, M. N. Baker, Elmer W. Firth, Myron S. Falk, Charles H. Haswell, W. H. Booth, H. A. Young and Rudolph Hering.

Rudolph Hering, M. Am. Soc. C. E.—The subject herewith presented for discussion embodies the treatment of the miscellaneous solid materials which are not removed by water carriage in underground channels, but require special cartage, and the best disposition of which is not yet generally recognized. The subject, in fact, is still under development, and, it seems to the speaker, has now reached a stage where it requires the assistance of the civil engineer to bring it to a satisfactory state of practice. Mr. Hering.

For convenience of discussion, perhaps also of treatment, the materials included may be divided into four classes:

I.—Ashes;
II.—Street sweepings;
III.—Dry rubbish;
IV.—Kitchen garbage, swill or slop.

---

* At the meetings of December 17th, 1902, and January 7th, 1903.

Mr. Hering. The problem before the engineer, therefore, is to effect a sanitary collection and disposal of these four classes of refuse at the least cost. For this purpose he must ascertain and carefully consider:

A.—The objectionable qualities of each class of refuse, viewed from a sanitary standpoint; which means the destruction of disease germs, the avoiding of a nuisance and the seeking of comfort and decency in the handling of this matter.

B.—The cost of collection from the place of origin to that of final disposal, and the cost of final disposal, modified by any credit that may be due to a salable value of any part or parts of the refuse.

A.—Sanitary Collection and Disposal.

I.—Ashes.—The objectionable feature in the collection and disposal of ashes is the dust. When placed on the sidewalk, when carted through the streets, or when dumped while the wind is brisk, the arising of dust is a nuisance.

The evident and simple remedy against dust, when collecting ashes, is the use of covered receptacles and covered carts, and, when disposing of them, is a careful and slow dumping, or, on special occasions, perhaps also a prior wetting. The average of city ashes has no manurial value.

The filling in of low ground is the usual way of disposing of this material permanently. Occasionally, it has been utilized as an ingredient for mortar or for road-making.

II.—Street Sweepings.—Street sweepings consist of horse droppings, accumulations of general city dust and refuse not separately removed, and earthy material worked to the surface of pervious pavements or brought by wheels from country roads.

The composition of sweepings, therefore, varies chiefly with the quantity and quality of traffic, the kind of pavement, and the distance from unpaved roads. The sweepings may contain a large or a small proportion of organic matter, some of which may break down rapidly and some slowly. The fineness of this material is its chief objection, as the wind readily takes it up and blows it about, to the injury of man, of beast, of food and of certain kinds of property. Germs of the most prevalent city disease, tuberculosis, may be abundant in the street dust, and the troubles of the continual warfare against this dust, when settling in the rooms of our houses, are known to all.

Sprinkling and sweeping at frequent intervals are the recognized, though often inefficiently executed, remedy against the dust nuisance within the city, and covered carts are also here required for the proper removal of the material.

Its final disposal can be effected in either of several ways. When mixed with or without fuel, and when containing but little inorganic matter, it can be burned, thus destroying effectually the organic matter, including disease germs, contained therein; or, it can be dumped for the filling of low land, which can be quite satisfactory in many cases. When dumped, the upper surface of the fill alone remains exposed to the wind, and this only temporarily until vegetation protects it. When street sweepings contain a large proportion of organic ingredients, as horse manure, they are utilized as a fertilizing agent, with more or less success, and with the concomitant elimination of their objectionable qualities. Mr. Hering.

III.—Dry Rubbish.—The dry rubbish of a community consists of that aggregation of material which is not classed especially under the other headings. It consists chiefly of wood, paper, straw, leaves, rags, leather, stoneware, glass, metals, and sweepings from buildings. Its most objectionable qualities are dust and lurking disease germs. It is not easily decomposed, and its odor, usually, is but slightly offensive. Its value consists partly in the material that may be picked out, assorted and sold, such as metals, glass, and other discarded articles of manufacture, and partly in the fuel constituents which it retains after such picking.

Its collection should be made in covered carts, to prevent odors, dust, and other nuisances. The frequency of collection is not an important question, due to the stable character of the material.

Its final disposition, after it has been picked over, is usually either as filling of land, by dumping, or as fuel, by cremation.

Filling up low land with this rubbish was an early practice, and still persists in many places. Where land is available, where such material will not cause future trouble, and where it cannot be burned economically, this practice cannot be seriously objectionable. Where the rubbish can be properly burned, such practice would generally be preferred by the medical profession.

IV.—Kitchen Garbage, otherwise called Swill or Slop.—Garbage proper is the refuse chiefly from kitchens, markets and slaughter houses, and consists almost wholly of rejected animal or vegetable matter in solid and liquid form. A large proportion, therefore, readily breaks down, and, on becoming putrescible, disperses offensive odors.

It is the most important part of city refuse, both because it is likely to create the greatest nuisance, and because it has the greatest inherent commercial value. It has been the subject of more discussion, its treatment has been more varied, and a satisfactory solution of the problem of disposal is less generally acknowledged, than in the case of the other classes of refuse.

Regarding the best manner of collecting this material, there is but little question. It is generally conceded that it should be placed in water-tight and covered receptacles, and collected in water-tight and covered carts or wagons. It is also conceded that the collection should be frequent. The frequency, however, varies often with the

Mr. Hering. density of population. Yet this density affects only the quantity of the material, not its liability to putrefy, which is controlled wholly by its composition, the length of time it is exposed, and by the temperature. That the latter fact is beginning to be recognized in some cities is evidenced by ordering more frequent collection in summer than in winter. The necessary frequency should also vary with the manner of disposal. If, for instance, garbage is fed to swine, then, for sanitary results, a daily collection is peremptory throughout the year, except only in very cold weather.

In northern climates, where winters are cold and long, the receptacles, for obvious reasons, are better made of wood than of metal, while in warm climates the reverse is true, which conclusion is already recognized occasionally. Where garbage is likely to freeze while waiting for a collection, a conical-shaped receptacle is the only proper form.

The final disposition of this material is now being effected in the following ways:

1.—Feeding to swine,
2.—Dumping as filling,
3.—Dumping into large volumes of water,
4.—Ploughing into soil,
5.—Extracting the grease,
6.—Cremating the organic matter.

Feeding to Swine.—This was one of the first methods of disposal ever used, and it is still practiced extensively. In New England it is the prevailing method in small and medium-sized communities. If kept free from other refuse, and from such as would injure its food quality, if collected and delivered properly daily, there seems to be no serious objection to this method, either from a hygienic or commercial standpoint. It has an advantage over all other methods in yielding a revenue from its sale, which is sometimes greater than that obtained through any other method of disposal. Its application, however, must be confined to conditions under which the sanitary quality of the food, respecting disease germs and odors, can be safely guarded and maintained, and which excludes it from adoption in large cities, except in the case of hotels and eating-houses, where a private collection can be secured.

Dumping as Filling.—This is practiced in places where feeding is found objectionable. When not mixed with other refuse, dumping is objectionable on account of the subsequent putrefaction, in case of a large accumulation of the garbage. When mixed with other refuse, the merits of dumping depend on the proportions of the ingredients of the mixture, and on the character of the locality where dumped. Under some conditions a nuisance would, and under others would not, result.

Dumping into Large Volumes of Water.—This method can, or can not, be advised according to the special conditions of the case. Like the disposal of sewage by dilution, a safe disposal of garbage can be effected by dispersion, if the quantity of water is sufficient, if the current or winds prevent a drift ashore, and if the sinking material does not accumulate so as to reduce the necessary channel depth for navigation, or produce foulness by deposit in shallow water. Opportunities for a satisfactory disposal in this manner will be rare. Mr. Hering.

Ploughing into Soil.—This has been practiced with varying success. If freed from metals and large entangling objects, and if mixed with barren sandy soils or acid soils, garbage being somewhat alkaline, good results have been obtained for the raising of crops under careful management. This process is at present considered favorably in several cities of Europe, because it can dispose of disease germs and odors satisfactorily, and because the gradual nitrification of the organic matter has produced a fertile soil where sterility had previously prevailed.

Extracting the Grease.—The extraction of grease from garbage is an American industry which has been promoted in a number of our cities. The process in its general form consists briefly in separating the material into two parts, each having a salable value, namely, oil, or fat, and a dry, somewhat stable, fibrous material of vegetable and animal origin, called tankage, and used as a filler or diluter for certain fertilizers. The separation is brought about by simple boiling and subsequent pressing out of the oil and water, which are in turn separated by gravity. Or, it is accomplished by an admixture with naphtha, which dissolves the oil, and later, after it is separated from the tankage and cooled, enables it to be readily extracted. The water separated from the oil is an offensive liquid, and the gases and fumes generated at the works are offensive vapors. Both require a careful and thorough treatment.

The objections to this reduction method of garbage disposal, which is still practiced in Boston, New York, Buffalo, Philadelphia, Washington, St. Louis, Cincinnati, etc., although abandoned in some cities, is due to the offensive odors occasionally, if not continually, arising from the same. The process requires the rejection at the works of tin cans and other foul material which interferes with the economical process of securing the oil. For such rejected material a different treatment and disposal is necessary. While, from an engineering point of view, it is practicable to construct and operate these reduction works so that odors will not prevail, the means for doing so are also expensive. As they are operated invariably as a business undertaking for profit, the conflicting interests naturally involve the constant danger of a possible nuisance. The practical result has been that, for the sake of sanitary security, such works are now placed as

Mr. Hering. far from inhabited territory as practicable, with the disadvantage that the distance for transportation has become correspondingly greater.

The advantages of this method lie chiefly in the commercial value of the oil and in the destruction of disease germs and putrescible matter by heat. There is little value in the tankage, which is used occasionally as a fuel or even wasted on dumps.

After all the operations of a reduction process have been conducted without offense, the financial aspects will decide the permanent merits and value of this method of disposal.

Cremating the Organic Matter.—This is the favorite and exclusive method of garbage disposal in England, where it has been used for fifty years; and it is also the favorite one of the medical profession. In the eyes of the sanitarian the complete destruction by fire of all matter that may become offensive, and at the same time of all disease germs that may lurk in the sweepings of the house and in the rejected food materials, is an ideal process.

In practice, however, the results have not always been ideal. In our country odors from such works have been complained of in many instances, and a number of crematories have been abandoned as nuisances. In England, however, such has not been the case. Furnace extensions are built every year. Complaints are rare. In Hamburg, Germany, where is the largest single garbage crematory in existence, this is giving no offense, although adjoining a built-up section.

An examination into the reasons for the different results achieved by cremation in Europe and America shows that the elements making up the problem are not the same.

1.—The material as it is burned in Europe has not the same composition as that which here has failed to burn properly. European success was achieved with refuse generally consisting of a mixture of garbage with ashes and dry combustible rubbish. The chief failures in America are recorded where garbage alone was cremated, without the admixture of other combustible dry refuse, and even after the addition of furnace coal. An examination into the combustibility of European and American combined refuse does not show a great difference. The few analyses we have show that cremation under proper precautions and suitable conditions should be as practicable here as in England, and recent furnaces built here demonstrate this fact.

2.—The essentials in the construction of the successful English furnaces differ from those of the furnaces peculiar to our country, where they have not been satisfactory. The English furnaces are built with sloping grates of small area, while the American furnaces generally have horizontal grates of large area. It is hardly necessary here to point out that in this element lies a main defect of our plants. A sloping grate admits of an almost automatic, continuous process of gradua drying, then heating, and finally burning; the fresh material

Mr. Hering.

being added at the upper end and the hottest fire being near the lower end. A horizontal grate, on the other hand, to secure proper combustion, depends almost wholly upon stoking. It causes the burning to be irregular, due to the sudden dropping of cold and perhaps wet material upon that which is already almost completely burned. In Europe we also find an extensive use of large dust chambers to prevent the escape of this fine material, of which we so often hear complaint. We find also a more frequent and effective placing of dampers to control the draughts, and thus prevent the escape of unburned, and therefore odorous, solid particles and gases. Finally, more attention is directed there than here to the evolution of the greatest possible heat for the incidental production of steam.

3.—The expert garbage fireman, who is considered essential to success in England, is generally supplanted here by a man whose only qualification for the position may be that he can shovel coal or pull out clinkers, but generally has not the remotest knowledge or even conception of the difficulties in burning, on a large scale, both continuously and completely, the most heterogeneous mixture of all forms of solid matter to be gathered from a modern community. The advantage of a trained expert in firing city refuse was well shown by the fact that until Hamburg imported English firemen to train the German firemen in the mere operation of their own plant, satisfactory results could not be obtained.

4.—In Europe, the municipalities, under the guidance of experienced engineer officers, generally undertake the cremation by their own long-trained employees, while here, usually, either the health boards or council committees select not only the design, but indicate the method of operation, appoint inexperienced employees, or enter into a contract for immediate profit rather than for permanent efficiency. The difficulties underlying the problem of city refuse disposal, which is almost wholly one of engineering, have been solved satisfactorily where competent engineers have been employed for the purpose. It is hardly to be expected that, without professional skill and training, desirable results can be reached in this, any more than in any other, branch of engineering.

It will therefore rest finally upon our profession to solve this urgent question in our country, by not only making the necessary preliminary investigations to determine the most economical means of disposal, to indicate the methods, and to design the works that will accomplish this in a sanitary and inoffensive way; but, also, by indicating the details of operation most suitable for the particular kinds of material delivered.

B.—Cost of Collection and Disposal.

In discussing this subject, in full view of what has been said regarding the sanitary demands, we are now able to add that the latter

Mr. Hering. demands can be satisfied by the engineer in any and all of their aspects if the necessary price is paid. He is therefore obliged to estimate, here, as in his other problems, the cost of each method, developed to a point of satisfactory efficiency. The least expensive of the several possible methods will then become the preferable one. As the question of cost embodies the collection, from the point of origin to the point of disposal, the estimate must embody the expense required between these two points; for one process may require a much longer haul than another; then, to be economical, the disposal in the first case must cost proportionately less than in the second.

Theoretically, it would be proper, in a given case, first to estimate the cost of properly collecting and disposing of each class of refuse, separately, in accordance with the several practicable ways indicated. Then it might appear that perhaps some two or three or all four classes of refuse could be treated economically by the same process. In this event a second estimate of cost would be appropriately made for a system of combining the collection and disposal of two or more such classes of refuse, and thereby the expediency of a system of separation or combination, in part or as a whole, would be ascertained. In practice, this procedure sometimes might be very simple.

In making such estimates of cost it is necessary to take account of both debit and credit sides of the financial aspect.

The collection is a debit in every case, its amount depending upon the distance the refuse is hauled. When garbage is fed to swine, the collection should be more frequent than in other cases, and a further debit would be proper on that account.

The final disposal has both debit and credit sides to be investigated. In the case of kitchen garbage and dry rubbish a substantial debit is required in establishing preventive and precautionary measures against the arising of offensive odors, of dust or other nuisances, in whatever process of disposal may be considered. On the other hand, the credits must be determined, as the case may be, and, if found to exist, they may be given to the several classes of refuse as follows:

Ashes.—By filling land, utilization for road-making, or as an ingredient of mortar, or otherwise.

Street Sweepings.—By making land or by utilization as manure, or otherwise.

Dry Rubbish.—By utilizing whatever value exists in the pickings, and in the remainder as fuel for power purposes.

Kitchen Garbage.—By utilizing its value either as food for swine, as grease and fertilizer, as a heat producer for power development, or otherwise.

If the subject under discussion is treated substantially as above outlined, the speaker believes: First, that, through the co-operation of

Mr. Hering.

the engineering profession, it will gradually cease to be a vexed question among our legislators; secondly, that the sanitary demands, in preventing nuisances and dangers to health, will be satisfied; thirdly, that the economical advantages inherent to the several processes will be secured; and fourthly, that the result will be an efficient process operated at the lowest cost.

Mr. Woodbury.

JOHN MCGAW WOODBURY, Esq.*—In a large city, all sorts of wastes are found: Wastes of ashes, garbage, street sweepings and rubbish; the rubbish consisting of paper, broken-down sofas, discarded mattresses, old shoes, bureaus, washstands, broken crockery, dolls, toothbrushes, tin cans, rags, bottles. Every conceivable kind of waste is either found in the streets, or is put out of the houses, for the city to dispose of. What shall be done with it, has been and is still the question. Heretofore, in New York City, most of these wastes have been disposed of by the archaic method of dumping them into the sea. This is but an extension of the first idea of disposal, which was to throw such wastes into the street.

This idea of final disposition of refuse still exists in less civilized communities, particularly in tropical countries. It is practiced all through the West Indies, the only sewer being an open one running through the center of the street, into which it is the general custom to throw all wastes, including offal. They have but little ashes, and the carrion crow and turkey buzzard are the common scavengers.

In New York City, at present, the speaker happens to be the common scavenger. In Paris, in the time of Louis XI, there was an order appointing the hangman as the common scavenger, and giving him permission to remove dead horses and cattle from the streets. The world has made much progress since that time.

The problem of garbage disposal is a vexing one. By the reduction process, phosphates and grease are obtained. By the incineration of garbage it is claimed that heat can be obtained, and that this will generate power; but all experiments in New York City show that it takes fuel to burn the garbage in a sanitary manner. The question of cost is important. The best or most scientific and sanitary process is not the best, financially.

Fig. 1, Plate XI, is a photograph of a flat scow being unloaded in the ocean, outside of Sandy Hook. It shows clearly the quantity of flotsam, much of which will be distributed along the beaches a little later.

Fig. 2, Plate XI, shows the Barney Dumper being unloaded at sea. It is a divided boat, or catamaran, and splits in the center, thus dropping its load. It is a great improvement over the flat scow.

The Delahanty Dumper, going to sea under its own steam, is shown

* Commissioner of Street Cleaning, New York City.

Mr. Woodbury. in Fig. 1, Plate XII. It is also a catamaran, having pockets hung between the two hulls. Doors, hung by hinges in the bottom of these pockets, are opened and closed by steam power, thus discharging the refuse. By the use of this vessel the refuse is disposed of at a cost of about 19 cents per cubic yard.

One of the dumping boards of New York City, with two Barney Dumpers and a flat scow, is shown in Fig. 2, Plate XII. This method of loading scows is also very archaic, as the ashes and light refuse are blown about by the wind. At these dumping boards, and on the scows while being loaded, a number of men are always engaged in picking over or sorting the refuse and trimming the scows. Quite a revenue is obtained by allowing the material of value to be picked from the waste, besides the advantage of having the scows properly loaded.

The output of rubbish and ashes from New York City is about 2 000 000 cu. yds. per annum. If the rubbish is separated from the ashes, it means a great deal, financially. At Riker's Island land valued at about $10 000 an acre is now being made by using the ashes as filling, which is certainly more economical than throwing this material into the sea and letting it go to absolute waste. Ashes make the best possible filling, and pack well behind the cribs, which are sunk on the mud bottom.

It is of the utmost importance that dumping at sea be stopped, not simply because it makes the beaches unsanitary and unsightly, but because it is a waste of valuable material. The separation of refuse is now in good condition. The quantity of garbage sent to the Sanitary Utilization Company in 1902 was 30 000 tons more than was ever sent in any previous year. This quantity of garbage has come mainly out of the ashes. The rubbish, also, is now kept separate in parts of the city, thus leaving the ashes almost a merchantable commodity.

For the disposal of rubbish Mr. Parsons has designed the incinerator on the pier at Forty-seventh Street and North River. It is burning about 60 loads of 5 yds. each, or about 30 tons in eight hours, every day. This rubbish consists of excelsior, mattresses, trunks, barrels, boxes, chairs, sofas, paper, rags, etc. These, in burning, produce sufficient heat to make steam in boilers, which will furnish power to run a belt-conveyer, now being constructed, which will be used for sorting the material. The surplus power will be used for lighting, for the elevator, for a traveling runway to the dumping board for loaded trucks, and possibly there will be some power which can be sold.

The City of New York now receives $110 000 a year for the privilege of sorting, and when greater facilities are given the sorters, as by the conveyer mentioned, a much larger sum will be realized.

Aside from the sanitary handling of the refuse, 63 acres of land have been made at Riker's Island, which is equivalent to $630 000; the

FIG. 1.—UNLOADING REFUSE FROM FLAT SCOW, OFF SANDY HOOK.

FIG. 2.—THE BARNEY DUMPER, BEING UNLOADED AT SEA.

privilege of picking over the refuse has been sold for $110 000 for the year; and the utilization of power from burning the non-marketable refuse will bring the total income up to about $900 000 a year. In this way the city's wastes become a source of revenue which will help materially in cleaning the streets. Mr. Woodbury.

H. DE B. PARSONS, M. Am. Soc. C. E.—The primary collection of the refuse of a municipality is made by carts, which convey the waste material to points for final disposition. This primary collection presents no special difficulties. It is a matter of carefully working out the details to suit best the existing conditions. Mr. Parsons.

The problem that is serious and difficult to solve is that of the final disposition of this refuse—a problem which is closely allied to the health of the community. The direct benefits derived by a suitable system for sanitary disposal are augmented indirectly by an increase in the health of the community.

No matter how the question is approached, the practical considerations of first cost and expense of operation and maintenance will be found paramount. On this account, as well as the short tenure of office of our city authorities, systems have come into use which are neither the most sanitary nor most economical for final disposition. Contracts have been let for such short periods that contractors have refrained from erecting suitable plants, and systems have been evolved which only earn for the cities a portion of the direct returns that might be obtained.

The refuse of a municipality consists of garbage, ashes, rubbish and street sweepings. In an average American city the divisions are proportioned approximately, as follows:

	By weight.	By volume.
Garbage	13 per cent.	18 per cent.
Ashes	80 "	57 "
Rubbish	7 "	25 "

The statistics gathered from one city are not applicable to another; and results predicated on information obtained from other places will not always be satisfactory. In order to make a final solution of the problem, it is necessary to study the conditions obtaining in the particular place under consideration. Garbage is combustible. It is extremely difficult, however, to burn pure garbage, and it has been found necessary to mix fuel with it in order to obtain complete combustion. The amount of fuel required varies from 8 to 20% of coal by weight, according to the character of the place and the constituent elements of the garbage. Ordinary kitchen garbage consists of approximately, by weight:

Animal and vegetable matter	20 per cent.
Grease	3 "
Water	70 "
Rubbish, cans, rags, etc	7 "

Mr. Parsons. Ash collections are divisible into two kinds, steam ash and household ash. The steam ash, produced by manufacturing companies, comes to a city dump practically pure. It is clean and contains little unburned coal. The household ash, on the contrary, contains a large quantity of unburned coal. It has been estimated that, in a city like New York, the quantity of unburned coal in the ash carted to the dumps from private houses and apartment houses amounts to about 20%, and, in the speaker's experience, there is little question but that the proportion is correct. This coal, mixed with the ash, can be and is utilized as fuel to assist in the combustion of garbage. The ash collections make a good filling material, and can be disposed of for that purpose.

Rubbish collections are very combustible, as are also the street sweepings.

Final disposition of the refuse may be made, first, by dumping on waste lands, or at sea; secondly, by sending the garbage to reduction works, cremating the rubbish, and utilizing the ash collections for filling; and, thirdly, by cremating the garbage and rubbish, and utilizing the ash collections for filling.

The first method is neither sanitary nor satisfactory.

The second method is sanitary, but not wholly satisfactory. The reduction works can only handle garbage, and the system is considered uncertain in case of an epidemic, as the employees might refuse to work the plant. Such works can only afford to operate on a bonus paid by the city. The process consists of extracting the moisture by pressure, as being cheaper than evaporation, and of cooking the material by steam in large tanks or digesters. The products are grease and a residue sold as fertilizer. The plants are expensive to construct.

The third method is sanitary, and, in the speaker's judgment, the best; because it has been accomplished without creating a nuisance, because it is the most sanitary, and because it can be made to produce the greatest revenue. To be successful, the combustion must take place at a high temperature, say between 1 500 and 2 000° Fahr. The household ash, the rubbish and the street sweepings will supply the fuel.

Cremation has been tried, in a number of places in America, with poor results, although very successfully practiced in Europe. The failure in American cities has been largely due to faulty design of the furnaces, lack of high temperature (a most essential feature), and want of means for properly caring for the garbage when first brought to the destructor. When garbage is allowed to collect in mass, it is most highly objectionable, and such collections have caused, in some instances, the failure of what otherwise might have been a successful plant. The heat generated can be used for various purposes and be made income-earning.

FIG. 1.—THE DELAHANTY DUMPER.

FIG. 2.—LOADING SCOWS WITH REFUSE.

CONDITIONS IN THE CITY OF NEW YORK. Mr. Parsons.

Area of Greater New York	320 sq. miles.	
" " Borough of Manhattan	39 "	"
Paved Streets, Greater New York	1 000 "	"
" " London	1 818 "	"
" " Paris	604 "	"
" " Berlin	500 "	"

The amount of refuse collected yearly in the Boroughs of Manhattan, Bronx and Brooklyn is about 2 800 000 tons. The collections, in pounds per capita per week day, are about as follows:

Garbage	0.4 lbs.
Ashes	2.5 "
Rubbish	0.3 "
Street sweepings	0.7 "
Total per capita per week day	3.9 "

Comparing this record with other cities, of which the records are believed to be fairly accurate, the following comparison is made:

New York	3.9 lbs. per capita per day.	
London	1.6 "	" "
Berlin	0.9 "	" "

Results varying so widely show the difficulty in predicting what can be done with a municipal refuse, unless one knows in advance the character of the waste.

The rubbish collections in New York City weigh, on the average, about 250 lbs. per cubic yard. In Boston they weigh about 202 lbs. per cubic yard.

These rubbish collections have been divided as follows:

	By weight.	By volume.
Combustible material	60 per cent.	37 per cent.
Incombustible material (valueless).	5 " "	3 " "
Marketable material	35 " "	60 " "

The rubbish is picked over at the dumps by a trimming contractor, who pays the city for the privilege. The value of this marketable material to the city is about $3.20 per ton. The Commissioner of Street Cleaning has stated that this figure is too low; probably it should be increased 50 per cent. It is figured as follows: The average rubbish collections are 300 tons per day, or 1 800 tons per week, of which the marketable or salable material is 35%, say 600 tons. For this the contractor pays the city approximately $1 920, or at the rate of $3.20 per ton.

The present Commissioner of the Department of Street Cleaning had an incinerating plant* constructed, from designs by the speaker,

* Illustrated in *Engineering News* and in *The Engineering Record*, in April, 1902.

Mr. Parsons at the city dump at the foot of West 47th Street, Borough of Manhattan.

The record of this incinerator is as follows: The number of loads of rubbish brought to the incinerator per day: Maximum, 80; minimum, 20; average, 44.2. The 44.2 loads per day weigh about 61 000 lbs., or 30.5 tons. Of this amount about 75%, or 22.87 tons, was consumed; 5%, or 1.5 tons, was incombustible; and about 20%, or 6.1 tons, was trimmed. Much more than 20% could have been trimmed, but the incinerator burned the material so fast that the contractor did not have time to separate it out.

The incinerator contains a grate surface of 90 sq. ft., and the stack is 120 ft. high. The average weight of the material consumed per hour is 5 720 lbs., or 63.5 lbs. per square foot of grate per hour.

The design of the incinerator, in brief, consists of three cells, each having 30 sq. ft. of grate area. The products of combustion pass over the cells into the smoke flue in such a manner that the product from Cell No. 1 has to pass over Cells Nos. 2 and 3; the product of combustion from Cell No. 2 mixes with the products from Cell No. 1 and together they pass over Cell No. 3; and that from Cell No. 3 mixes with the products from Cells Nos. 1 and 2 and is intimately mixed again in passing along a tortuous flue to the base of the stack. The result of this arrangement has been highly satisfactory, as regards the non-production of smoke. Taking a stormy day when the material was brought to the incinerator wet, the smoke was seldom visible for more than about 75 ft. from the top of the stack, and then only during the period of stoking one of the grates.

This incinerating plant was constructed as an experiment. In order that it might be free from any hindrance by injunction or otherwise, lest it create a nuisance to neighboring property, it was decided to locate the plant on one of the city piers, about 250 ft. from the bulkhead line. The permanency of location, of course, was not considered, the idea being that if the plant could be constructed quickly and show that combustion could be carried on without creating a nuisance, it would lead to the introduction in the future of other stations, more favorably situated, and at which better facilities could be provided for the reception of the material and for the picking of the same.

The incinerator, thus far, has proved satisfactory, and the Commissioner has authorized the preparation of plans for further developing its capacity.

An analysis of the cost of incineration compared with removal of material by scows is as follows:

First, the Final Disposition by Scows, as Formerly Practiced.—Assuming the same figures as now obtained at the incinerator, namely, 30½ tons of material per day, on the average, the cost of removal of that material, figured at 15 cents for towing, and two days' hire of the scows (one for loading and one for towing) at 23 cents, comes to about 38 cents per ton. Owing to the bulkiness of rubbish, only about one-half tonnage can be put upon a scow, the tonnage being figured by volume, which makes the disposal of the material come to something like 80 cents or more per ton by weight. Therefore, 24.39 tons at 80 cents cost the city $19.51 for removal. The 6.1 tons, making up the 30.5 tons, which is saved by the trimmers, at $3.20, also figures $19.52; thus, practically, the cost of removal by scows was offset by the receipts from trimming. Mr. Parsons.

Second, by Incineration.—The result is approximately as follows: Received from trimming 6.1 tons at $3.20, $19.52 per day. Received from the city ash collections, which at this dump amount to about 600 tons per day, and which are made salable at, say, 5 cents per ton when freed from the rubbish, $30.00. Total receipts to the city, $49.52. The cost to remove the incombustible material, 1.5 tons, at 38 cents, 58 cents; cost of burning, interest and repairs on the incinerator, estimated, at $7 per day; labor, feeding and firing, four men, $10 a day; ash removal, $1 a day; total, $18, Net gain, therefore, to the city, $30.94 per day; yearly gain, more than $9 000.

The incinerator, at these figures, will pay for itself within two years. According to the figures quoted by the Commissioner, namely, 19 cents a ton for the ash collections, would make a daily revenue of $114 from the ashes, in place of $30, or a net gain of $114.94 per day, or more than $34 000 per annum; at which rate the incinerator would pay for itself in less than six months.

The ashes produced by the incinerator are very light, completely burned, and, owing to the large quantity of wood from which they are produced, have some value as a fertilizer, and are used by the Park Department for spreading on lawns and around trees.

The improvements referred to are an unloading floor and a belt conveyer, which should increase the amount of material trimmed, that is, the marketable material; and should reduce the cost of labor. Probably the services of two of the men now required could then be dispensed with.

The problem before a City Commissioner—to dispose of the refuse of a municipality—is not always a scientific one. Usually, he has to face a practical problem. In New York City, the fault is due to the form of government and the contract system now in vogue. As a change cannot be made quickly, one has to do the best he can.

The scientific treatments of the wastes of a municipality are fairly well understood. The question is one of engineering; and an engineer who is willing to study the problem can dispose of those wastes in a manner which will be sanitary, provided, of course, that the munici-

Mr. Parsons. pality will give him, or will give the Commissioner, the money to carry out his plans.

The wastes are being disposed of in Europe in a more scientific manner than in this country. The reason, probably, is that action was taken in America without previous scientific study. In Europe, they started with cremation. In America, they followed with cremation; but, as the cremation system adopted was false, failure followed, and, in many localities, public sentiment was set against it. This has been probably the chief reason that cremation is not more generally adopted on this side of the Atlantic.

In Europe, the collections are made and brought to a destructor, where they are cremated; and, in the most successful plants, they are cremated without the use of any fuel whatsoever, beyond the fuel supplied in the garbage and by the other wastes. Each collection district differs in the amount and character of its refuse, but there is sufficient combustible material collected which, when mixed with the garbage, will make it burn in a satisfactory manner.

The object sought, in the most successful plants, is to produce a high temperature; and a plant that will not produce a high temperature is not considered successful.

An incinerator or crematory, as the destructors are called in this country, which has to have a fume or stench cremator, is false in design, or else has that element put into the design in order to meet public sentiment. If the temperature in the burning furnace is sufficiently high, there is no need of a fume cremator. A fume cremator is simply a small fire placed either at the base of the stack, or somewhere in the passage of the gases, the idea being to burn up any fume-creating particles that might pass from the furnace before they reach the outside air.

In Europe, recent practice has not adopted a high stack, many successful furnaces being operated with a short stack and a forced draught. By a short stack is meant one of 70 to 80 ft., although some are still lower. Sufficient height should be given the stack to stop any question as to its being a source of annoyance or a nuisance. The earliest destructors used very tall stacks and natural draught.

Personally, the speaker favors the destruction of municipal refuse by cremation, and believes that it will be generally adopted in this country before a great many years have passed. As indicated before, persons have promoted furnaces, destructors, cremators and incinerators which were not designed scientifically, or have formed companies not strong enough, financially, to carry out their claims. Several cities have recorded failure from both these causes, and in some the destructors are now out of use.

There has been another cause of failure, due to the collection of garbage at the destructor. Garbage, when allowed to collect *en masse*, is extremely offensive; and, unless it is properly taken care of and promptly burned, is apt to create an adverse public sentiment, although the destructor itself may be good of its kind. Mr. Parsons.

The system of reduction has been adopted by a number of our larger cities because it was easier to organize a reduction plant than a destructor plant. The fear of creating a nuisance—one of the causes most frequently mentioned—has acted to delay progress in the direction of cremation. The same cause, however, was not as active in the case of reduction. The returns from a reduction plant are not sufficient to operate it. Such a plant must rely upon a bonus from the city. The returns from properly constructed destructors, supplied with suitable municipal wastes, should enable the plant to be more than self-maintaining, and to earn a handsome income for the municipality.

It is quite common in Europe to see small towns in which the actual pumping of the water supply is accomplished by means of the steam generated from the heat produced in the destructors. Some towns use the power for electric street-lighting, and others sell the electricity generated. It is perfectly possible to use the heat generated in a manner which will be income-earning, while the reduction plant needs to be sustained by a bonus.

Destructor plants will probably first be started by the smaller towns. On their success, which depends on proper engineering advice, the larger cities will slowly plan future improvements.

The practical problem in any city is the distance from the point of collection to the point of discharge of the carts. In some cities that haul is short, in others it is long. In New York City, some of the carts, the speaker believes, have hauls as long as five miles. A system of collection by automobiles has been discussed, but no automobile has yet been designed which would be sufficiently regular to stand the service. Perhaps, in time, such perfection may be reached as to warrant the collection by this means.

Snow-melting machines cannot handle the snow fast enough. If the carts could be regulated so that they would bring the supply just at the rated capacity of the machines, they might operate in a satisfactory manner; but the trouble is that all the carts arrive together. As they cannot stand in line with their loads of snow, they have to dump it, and if the machine cannot take it it is simply dumped on the street, which is just what is not wanted. When snow has lain in the street a little while, it becomes mixed with dirt and is then difficult to melt. Snow-melters have not proved economical, either in time or money.

If the snow is collected from the streets of New York City promptly, there is little objection to its being dumped in the river, as the snow is very clean. A great drawback has been the lack of docks from which snow could be dumped.

Mr. Parsons. While Mr. Soper's remarks are interesting, some of his statements scarcely give due weight to the practical conditions as they exist in the City of New York.

The speaker is sure that anyone studying the conditions as they exist in this city would conclude that the problem is less scientific than practical. In order to turn the practical problem into a scientific one requires time, and in time it will probably be done.

What all cities and all municipalities need is a co-operation between the different departments. The present Commissioner of the Department of Street Cleaning has done more than any other Commissioner in this direction. A Commissioner is hampered by the General City Government, of which he is but a part. The failure of some of the former administrations, to clean the streets and dispose of the refuse properly, may not have been due solely to the Commissioners.

No criticism of Mr. Soper's purpose is intended. His remarks seem to be misleading to members who have not made a specific study of the subject. His criticism of some of the former Commissioners is a little severe. While the results may have been as stated, it is possible that the Commissioners did not have the power to remedy those conditions.

The garbage of New York City contains an enormous amount of water, and the worst class of all to burn is the stuff collected from the fruit markets, as it is practically non-combustible, except by strong aids. Ordinary offal or kitchen garbage will burn reasonably well when mixed with the other wastes.

The rubbish collections in New York City are 95% combustible; the ashes from the private houses and small apartment houses contain about 20% of unburned coal. If the rubbish collections and the ash collections from private houses and small apartment houses are mixed with the garbage, the whole can be burned in a suitable furnace.

The ashes from the steam-heating companies and large manufacturing plants, known as "steam ash," are collected and brought to the dumps in separate vehicles, so that there is no necessity at any time of ever passing such ash through a destructor, because that ash is practically pure.

In Europe, the principal system of collection is one of non-separation. They simply collect the whole of the refuse, and this, just as it comes, goes into the destructor.

Most destructors contain some form of what is called a "drying table." The material as collected, either wet by contained moisture, or wet by rainfall, is placed first upon some part of the cell or furnace where it can be warmed, and is gradually passed back over the fire, in a manner similar to the coking method of firing. When the material has become dry, it is forced back to the hottest part of the grate, where it burns rapidly, supplying heat to dry the incoming material, while sufficient heat is left to evaporate, in boilers, according to the material burned, from 1 to 2 lbs. of water per pound of refuse. Mr. Parsons.

W. F. Morse, Esq.—For twelve years the speaker has been actively engaged in the work of final disposal of municipal refuse. It has been his province to present the question in many places and in many ways, and, until now, it must be candidly said that it has not met with appreciation or attention on the part of the engineering fraternity. It has been a question which has fought its way, despite opposition, has been in the hands of men and committees who were really unacquainted with the subject, and, consequently, has been approached from the standpoint of an endeavor to solve an engineering question without the benefit of the best engineering skill. Mr. Morse.

The whole subject is a very wide one; it cannot be treated in a short way—by a short discussion—it has to be taken piecemeal and dissected, each part separated and treated by itself.

The remarks of Mr. Hering cover the whole question in a general way. He has given an outline of the various methods of collection, of the disposal, and of the results obtained. The remarks of the Commissioner of Street Cleaning and of Mr. Parsons cover the work done in New York, but there are many other cities than New York investigating this matter, and it is of much wider scope than the limited field of a single city.

In the limited time allowed it is possible to deal with the question only in a very general way, and to give a brief sketch of what has been done in the past ten or twelve years in this country, and, in comparison with that, an idea of what has been accomplished abroad.

Municipal waste has been divided into four classes: garbage, street sweepings, ashes and refuse. The disposal of all these is accomplished by two methods, or systems: reduction and cremation. Reduction, or the treatment of garbage only for obtaining a commercial product, is not now under discussion; but it may be said in passing that this method has been adopted by several of the larger cities, has cost large sums of money to establish, and is yet at a stage where it is impossible to say what its future will be. It is only known that companies and men are putting large sums of money into reduction processes, and that, so far, as many plants have failed and have been abandoned as have succeeded.

If a city is investigating this subject with a view to the disposal of all its waste, and if it adopts a means of disposal which treats only one-tenth of it, there is manifestly left on its hands the remaining nine-tenths. Reduction deals with only 10 or 12% of the whole collected waste of the city, treating garbage only. Cremation, on the other hand, takes the whole collection of miscellaneous waste and disposes of it all.

Mr. Morse. There have been in the United States, in the last twelve or thirteen years, something like seventy-five or eighty places where cremation has been adopted. This work was begun in Des Moines, Iowa, about fourteen years ago, by a furnace invented by Mr. Andrew Engle, and from which have come all the forms of the American cremator, all more or less imitations of the original Engle type. It is a structure of rectangular form, having an upper chamber and transverse bars on which the garbage is deposited. The fire is at one end, and in most cases the secondary fire for burning the gases is at the other end. In all cases of American crematories two fires have been found necessary, and that principle has been maintained in constructions of this type.

About a month ago, an article appeared in one of the prominent municipal journals under this caption: "Why do American Crematories Fail?" The text of that article was a report, made by a leading engineer of this Society, upon a furnace built in a neighboring city, and which was reported to be working unsatisfactorily. In the main essentials, that article is true, and, after twelve years' experience in building garbage furnaces in this country, the speaker is in a position to say that satisfactory progress has not been made; that while one or another type of furnace after the American pattern has been built and has performed its functions, and may be still doing so, the retrogression—that is to say the abandonment of furnaces which have been built after these patterns—has kept pace with the number installed. Therefore, comments, by the newspapers and by prominent engineers, to the effect that the progress of the disposition of waste in this country is not satisfactory, are absolutely true.

Beginning fourteen years ago, with that form of American cremator which was shortly after imitated by other constructions, there have been seventy-five built, and of these perhaps the most widely known and successful installation of the American type was made at the World's Fair in Chicago. This was a case where it was expected that some 50 or 60 tons of refuse per day would be handled, also the sewage sludge which could not be disposed of in any other way. Furnaces, after an improved Engel pattern, were built, and were run six months without intermission, using oil as fuel, and destroying on the grounds of the Exposition, within the confines of one building, all the garbage and all the sewage sludge caused by the presence of 27 500 000 persons, without sanitary nuisance, and to the entire satisfaction of the authorities.

That is an example of what might be done under the best conditions in which an American furnace could work, but that example was never improved upon. The furnaces so built were offered as a gift to the City of Chicago, but they were declined on the ground that they were placed on public property, under control of the State, not the City, and could not be accepted by the City. The results of that demonstration in Chicago show that under proper conditions garbage containing a large percentage of moisture and sewage sludge (two of the most refractory things in the world to deal with by fire) could be destroyed by maintaining a sufficiently high temperature and providing combustion chambers for the destruction of the gases. Unfortunately, this example was never carried much further by that company, but it was acted upon by others, and has been repeated since; yet, in spite of the lessons taught by ten or twelve years of experiment, it may be said that the progress made in cremation work has been equalled by the retrogression in the same line; that furnaces are abandoned about as often as they are built. Mr. Morse.

There are several obvious reasons for this. The work has been begun at the wrong end. At the outset, a sufficient amount of engineering skill was never applied to it. Municipal committees have gone about the country and have been persuaded by ambitious furnace builders to install plants. These plants have been simply experimental—built for a profit. They were not durable, and required extensive repairs. In many cases they were not sanitary; they emitted odors; they were not in all respects adapted to the work required of them, and they were expensive to operate. Yet, with all their defects, they have made for progress in one particular: they have directed the attention of the people to the fact that something is needed.

At the present time there are four or five companies constructing garbage crematories with varying success in different parts of the country. There are at least ten or twelve firms who have gone down—in other words, who have constructed furnaces found to be inefficient—and we are at a stage of the general question of cremation where it is necessary to have some more efficient means brought forward. If progress is to be made, it must be made either by inventing a new furnace, or by adopting other furnaces which have been proven successful elsewhere.

The question arose in England twenty-five years ago, long before it was ever thought of here. It began by the construction of the cell furnace, so-called, the Fryer destructor, operating by the use of a natural chimney draught, no forced draught being thought necessary at that time. The results were almost identical with those of the American furnaces of to-day. The temperature was low, offensive gases were thrown off, notwithstanding the "fume-cremator," or secondary fire, and it was found expensive to maintain. No real progress was made until about four or five years ago, when the furnace was shortened, forced draught introduced, the sloping platform made in such a shape that the garbage could be used as part of the fuel, and a combustion chamber added, behind which a steam boiler was placed.

The English destructors are to-day one of the most powerful agencies for sanitary conditions in England. They take what is

Mr. Morse. known here as municipal waste, as a whole. No separation is made; the entire waste of the household goes into what is called an ashpit, or ashbin—garbage, refuse and ashes—and this is taken directly to the destructor and burned. It is destroyed under forced draught, with no other fuel needed than the material itself. The result is the production of a volume of heat which operates steam boilers of very great capacity, the power from them being utilized for the production of electricity, for pumping sewage and water, and for many other municipal purposes.

The comparison between English and American efficient disposal work is distinctly against us. We are doing as they did six or eight years ago. At the present time there is no English destructor on this continent. The Thackeray furnace is an attempt. It is really an imitation of the principles of the Fryer destructor which was built many years ago in England, and which has now almost passed out of date, being supplanted by more powerful ones. There are two examples of this furnace, one in Montreal and another in San Francisco; but in the last five years, since these furnaces were built, no others have been installed, and there has been little attempt to develop more fully the English ideas.

The chief reasons for this lie practically on the financial side of the question. The cost of the construction of a destructor is about one-half more than the cost of an American cremator of equal capacity, and municipalities are usually looking first to the dollar and second to the sanitary operation and to the comfort of the people.

The following figures are quoted from the latest information regarding the operation of the English destructors. The comparative results from twenty English destructors, of eight different types, with thirty-two tests made for evaporation at different periods, shows that the average combustion per square foot of grate surface is 47 lbs. of mixed waste, garbage, ashes and refuse, and the evaporation of 1.36 lbs. of water for every pound of waste consumed. Therein lies the secret of the progress made in England. When they put into their destructors a mixed mass of material, burn it under a forced draught, and develop steam power which is equivalent to the evaporation of $1\frac{1}{4}$ lbs. of water for every pound destroyed, the enormous force produced can readily be seen, the utilization of which has made it possible for every municipality to do its work with its own power, without other fuel than the material destroyed, at a very small cost as compared with the cost of operation on this side of the water.

The cost of labor for destroying a ton of mixed refuse will run from 12 to 30 cents per ton, dependent upon local conditions. In one place the work is carried on at a profit; in others, the returns from the destructor, in steam power and in products manufactured from ashes, pay all operating expenses and fixed charges. The residuum from these furnaces is a valuable asset. The clinker is ground up and made into mortar, or tiles, paving blocks and brick. Mr. Morse.

There is another item on the credit side of the destructor, that is the utilization of power developed by the combustion of municipal waste for use in electricity works.

There are about 160 places in Great Britain in which destructors are used, and more than two-thirds of these utilize the power for various purposes. The electric force developed by a destructor which burns the waste products of a city is equivalent to 30 British Board of Trade Units per ton of waste destroyed. If it be assumed that each unit is equivalent to 280 candle-power, the result is something like 8 400 or 8 500 candle-power from each ton of waste.

The chief reason why advance in disposal work in that country has been so rapid and efficient is the fact that English municipalities proceed upon a different plan from ours. They make it a part of their municipal work. In this country it is contract work, and, consequently, there is not the same opportunity for cities to develop along the lines of municipal ownership and administration. There is a strong desire on the part of municipalities to adopt these improved methods, but they are held back by the fact that in almost every case the contract takes precedence, and the contractor is desirous only of getting through with his contract, and the same results are not accomplished as when the work is done by the municipality.

The remarks made by the Commissioner of Street Cleaning and by Mr. Parsons upon the work in New York City are pertinent and apropos of the question, but they do not cover the country. The principles which Dr. Woodbury is carrying out were instituted by Colonel Waring five or six years ago, and the experiments made at that time showed conclusively that the value of the rubbish of New York City was greater by far than anyone had supposed. The statement made by Colonel Waring in one of his last reports, that the value of the refuse, garbage and ashes of New York City would, if properly utilized, pay one-half of the total expense of cleaning the city, has never been realized. Yet it is certainly a question which is being answered by the present administration in a very satisfactory way. The Commissioner has alluded to, and Mr. Parsons has described, an incinerator, for the conversion of refuse, at West Forty-seventh Street.

The speaker had the honor of presenting this question to the Mayor of Boston four and a half years ago, proceeding largely upon the lines marked out by Colonel Waring, and at that time it was believed that the utilization of the refuse would become a feature of very great importance in municipal work. Mayor Quincy looked into the matter personally. He called for specifications and bids; he obtained four or five bids which were unsatisfactory, and rejected them. He took the plans which were offered for the construction of an incinerator in Boston,

Mr. Morse. and caused it to be erected by a private company. The incinerator is quite different in pattern from the one described by Mr. Parsons, but for four years it has been burning the refuse of the business section and a large part of the residential portion of the city, taking the waste from about 200 000 people, covering about 95 miles of streets, brought to one central point. The waste is carried into a room and dumped upon a moving platform, or belt, 150 ft. long, and during its passage the refuse is picked over by men placed on each side. These sorted portions pass into bins and presses below, and the remainder is discharged automatically into a destructor, or incinerator, placed across the end of the building. The quantity of waste destroyed began at about 25 tons per day for the first year, and it has increased until now about 40 tons per day are collected and disposed of. The product which is picked out from the refuse by these men will reach very nearly 60% of the whole bulk. It is divided into six or seven classes of paper, five kinds of rags, rubber, glass and iron. Everything possible is sorted out and saved, and the residuum destroyed are the small, worthless portions of paper, excelsior and straw, and a small amount of wood. The operation is conducted precisely as it should be conducted in New York City, in a closed building, the city carts discharging their loads out of sight of passers-by, no nuisance of any sort being connected with the work. After baling, the salable portions are at once conveyed to market, and the power produced by the combustion of the worthless portions of the refuse drives a 60 H.-P. boiler to its utmost capacity, which in turn moves all the machinery, lights the building with electric light, removes the ashes, heats the building in winter, and, thus far, has furnished a surplus of steam power.

The operation of this plant for four years shows that the refuse of a city of the size of Boston is more valuable than almost any other part of the municipal waste, and that every city with a population of 50 000 or upward can easily turn its refuse into money which will go a long way toward paying the expenses of the whole collection service.

Other cities are considering this question, with the idea of bringing into use a system similar to that which has been carried on in Boston, and which has just now been reinaugurated by the Commissioner of Street Cleaning in New York City, and which will undoubtedly be a factor in the disposal work of all cities in the country.

The average quantity of refuse collected is about 1 ton per day for every 10 000 persons. It may be more or less according to the different character of the city and the population, but it is of sufficient value to pay for the installation and operation of a plant in any one of the larger places in the country.

The future of this question depends upon utilization. American crematories have been going on for ten years, destroying garbage by using coal. The builders, up to this time, have paid no attention to refuse and its salable and fuel qualities. They have allowed the heat produced by the combustion to go to waste; and they have not been over-desirous of guarding against sanitary nuisance, but have sought to destroy the waste in the shortest possible time, without considering the economical points involved. Mr. Morse.

There is a general impression that a change must be made before real progress can be had, and there is hope that the engineering profession throughout the country will take up this question, as has been done abroad, and bring its trained experience to the assistance of the municipalities, solving the question in a much shorter and easier way than it can otherwise be done. It is inevitable that this will come about. The men who are trained in constructive and operative work should be the ones to decide what shall be done in this line by every city.

There is another side to this matter which is pertinent to the question under discussion. That is the disposal of the waste of large isolated buildings, like public institutions of every class, hotels, business establishments and even individual dwellings. The quantity of waste produced in these places is proportionately larger than under other conditions. There is now in use a form of destructor, installed usually in the basement of the building, connected with the boiler breeching, or the chimney. These destructors are used for the disposition of every class of waste, including that from laboratories and medical colleges.

The latest examples of these furnaces have been built in connection with the steam boilers of large business establishments, where the heat from the enormous mass of waste destroyed is used as auxiliary for operating the boilers. In some cases the destructor is also provided with a separate steam boiler operated by the furnace, but supplied with its own fire-box, the power from which is utilized for specific purposes, like running elevators, or in connection with the low-pressure steam system of the building.

The advantages to be obtained by the introduction of these local destructors are the disposition of the waste on the premises in a perfectly sanitary manner without the necessity of storing and handling the same, and the development of a certain amount of steam power, resulting in the saving of coal, or other fuel, and to some extent expense in steam boiler equipment. These destructors are also being adopted by the Government for use in Army and Navy posts and government buildings throughout the country.

The number of these destructors in use, already quite large, is increasing rapidly, as the installations are made under the direct inspection and by the advice of competent architects and engineers engaged in the construction work.

Mr. Soper. GEORGE A. SOPER, Assoc. M. Am. Soc. C. E.—The question of the sanitary disposition of municipal wastes is often divided into two parts: Final disposition, and collection.

By final disposition is meant the utilization or simple deposit of wastes; and by collection is meant all other steps in the operations of a street-cleaning department. It is not uncommon, in discussions on this topic, to omit that part of the subject which relates to collection; but, as the two questions are inseparably related in practice, it seems desirable that they should be treated jointly.

To a large extent, the method chosen by a city for the final disposition of its refuse is a controlling factor in the organization and plant of its street-cleaning department. This has been so in New York City.

The history and development of the Department of Street Cleaning in New York City affords many interesting suggestions as to the relations between final disposition and collection, and the experience of that city furnishes, withal, such an instructive illustration of the need of keeping the various parts of the machinery of the street-cleaning department in harmony and up to the requirements of a rapidly growing municipality, that some extended reference to it may be useful.

The organization of the Street-Cleaning Department, as a separate branch of administration in New York City, dates from the year 1881, up to which time the collection and disposal of wastes was carried on under the direction of the Department of Police. The new organization was the outcome of much agitation on the part of the public; physicians and political reform associations taking an active part in the crusade.

For ten years after its formation the Department did little to justify its existence. On the contrary, it became such an object of dissatisfaction that, in 1891, it was determined to appoint an Advisory Commission to inquire into the defects of the organization and to recommend measures of improvement.

The Commission was named by Mayor Hugh J. Grant, and consisted of five citizens, among whom were two business men, a lawyer, a sanitary expert and an engineer, the last being a member of this Society.

The report of Mayor Grant's Commission marked an epoch in the history of the Street-Cleaning Department of New York City. It is, moreover, so remarkable a document, from the clear and exact way in which the faults and needs of street-cleaning departments in general are defined, that a summary of its findings is here given:

In the first place it was declared that the department was not efficiently managed, as judged by standards of management of well-established corporations engaged in large enterprises.

One source of difficulty lay in the methods of employing labor. The laborers themselves were inefficient, and owed their positions to political influence. Mr. Soper.

The plant of the Department was poorly located.

The plant was insufficient.

There was a lack of proper organization and disposition of forces.

The manner of disposing of refuse was not satisfactory.

The law under which the Street-Cleaning Department was conducted was adequate, provided it was efficiently administered by a Commissioner who had the cordial co-operation of the Mayor, Police Justices, Department of Health, Commissioner of Public Works, and Board of Estimate and Apportionment.

The annual appropriation was insufficient.

The police should help to keep the streets clean.

There should be rules for the employment of men.

It was observed that the management of the Street-Cleaning Department required the same kind of ability as the management of a transportation company, or other large corporation employing a considerable number of men.

The report of Mayor Grant's Commission, one of the most forcible and lucid papers on this topic which the speaker has ever read, is the foundation upon which the present organization of the Street-Cleaning Department is based. The practical effect which the report had was a reorganization of the Department by legislative enactment in 1892.

Unfortunately, the opportunities thus opened for developing an efficient system of street cleaning were not at first realized. The first thing done in accordance with the recommendations was to increase the appropriations. Then progress ceased. The Department promised to remain as it had always been, an organization in which political influence was carefully and continually fostered.

By the time Mayor Strong's reform administration of 1895 came into power it was apparent that there was need of a strong head to develop the possibilities of improvement contemplated in the legislative act of reorganization. It is now a subject of universal satisfaction that such a head was found by Mayor Strong when, in 1895, he appointed the late Colonel George E. Waring, Jr., Commissioner of Street Cleaning. Whatever may be said of Waring's career, as an engineer or sanitary reformer, it must be admitted that in the three years during which he held office as Commissioner of Street Cleaning he made comparatively few mistakes and reached the greatest success, in raising the standard of usefulness of his Department, that New York City has ever known.

At the same time, it is not to be denied that circumstances favored Waring. A large part of his success was achieved on lines laid down

Mr. Soper. in the report of Mayor Grant's Commission of 1891. Waring also had the advantage of a report of a Legislative Committee of investigation and advice appointed by Mayor Gilroy in 1894. The report of Mayor Gilroy's Committee, dated November, 1894, recommends that:

First.—Dumping city refuse in the harbor or its tributary waters should be absolutely prohibited.

Second.—The ordinances requiring householders to keep separate the garbage, or kitchen refuse, from ashes and other house refuse should be rigidly enforced.

Third.—All house refuse should be collected in galvanized iron vessels with tight-fitting metal covers, of such size that, when full, they can be handled by one man.

Fourth.—Daily collections of garbage should be made by the City and delivered at dumping wharves into temporary storage, or self-propelled boats of some approved type to be furnished by the party having the contract with the City for the final disposition of the garbage.

Fifth.—The garbage should be disposed of by a reduction process producing fertilizer and commercial grease; and the City should invite competition by the various companies controlling such systems, in order that the greatest benefit to the City may result.

Sixth.—A separate collection should be made of the remainder of the refuse of the City "not otherwise provided for," which should be used for filling Riker's Island, or elsewhere; and, for transportation of this material, a sufficient number of self-propelled boats of approved type should be constructed and owned by the City. It should be provided, however, that whenever the whole, or any considerable portion, of the street sweepings can be disposed of by contract or sale for fertilizing purposes, at a price greater to the City than the value for filling, then such disposition should undoubtedly be made of them.

A careful reading of the foregoing recommendations will reveal on the one hand some excellent advice and on the other some stipulations which had the effect of committing the City to a particular process of garbage destruction and the plan of disposing of garbage by contract with a private company.

Some prominent reforms introduced by Waring were: Improved methods of administration; insistence that the laws and ordinances relating to cleanliness and order in the streets be enforced; and improvements in methods of collecting and disposing of the wastes. A large part of Waring's success, as the head of the Street-Cleaning Department, was due to his remarkable power of instilling a spirit of industry among his men. It has been well said that he found the force of sweepers and drivers a laughing-stock and left them a body of honored workmen.

It is well known that, after reorganizing his working force, Waring's ambition was to devise means for the economical, if not profitable, final disposition of the City's refuse. Up to Waring's time, the ashes, paper, garbage, and other refuse of the city had been collected and carried in a mixed condition to points along the water front where it was picked over by Italians, who had the privilege of extracting bones, bottles, rags, metals, and other salable matter. This privilege was purchased from the City annually, and in 1894 amounted to \$140 000. The residue left over after the picking was taken by scows and dumped at sea or used to fill in low lands in the neighborhood of the City. Mr. Soper.

Under Waring, the City wastes were separated, as is the custom to-day, into three classes before collection; notably, garbage, ashes and paper. Experimental studies were made of the value of these different kinds of refuse. Garbage was also collected from other cities, and the value of its grease and other constituents carefully determined. According to estimates, made at the time of these studies, the average quality of garbage produced in New York City was worth \$2.47 per ton.

A practical plant for the separation of the utilizable ingredients of refuse, other than garbage, was erected and operated for many months. The sale of the bottles, metals, rags, paper, and other useful materials, thus collected from 31 300 tons of refuse, netted the City \$3 680. A considerable amount of useful steam-power was also generated at the same time.

It is a lasting source of regret that Waring's term of office was too short to enable him to complete the development of a plant whereby all the wastes of the City could be disposed of profitably, or at least inoffensively. Just as his investigations were pointing to positive and permanent reforms his term ceased.

At the end of Waring's administration the efficiency of the Department immediately fell. The Commissioners who followed seem not to have been suited by talents or experience to conduct the Department on the plane of efficiency which Waring had initiated. Negligence, extravagance—some have said, corrupt practices—crept into the affairs of the Department. A single illustration will suffice to convey the speaker's meaning. The following extract is taken from page 24 of the Annual Report of the Department of Street Cleaning for 1900:

"In September, the New York Sanitary Utilization Company complained that it was not receiving all the garbage to which it was entitled by the terms of its contract with the City, on account of faulty separation, and claimed that the scow-trimming contractor was culling bones and grease for sale to which they, the New York S. U. Co., were entitled, and to which the scow-trimming contractor, by the terms of his contract with the City, expressly denied any claim or right. The matter was finally compromised by the Department paying to the scow-trimming contractor sixty cents per barrel for bones and one and three-

Mr. Soper. quarters cents per pound for grease collected from the ashes, which was turned over to the Sanitary Company. This arrangement resulted in a continual increase week by week in the quantity of this material collected from the ashes and street sweepings, so that for the last week in December the Department's payments to the scow-trimming contractor amounted to $2 127.41."

The charter of Greater New York, which went into effect in 1898, added much to the responsibility of the Street-Cleaning Commissioner, and complicated the unsolved problems of his imperfectly developed department. The plant of the Department should have been increased and the organization properly expanded to meet the requirements of the Greater City, but it is clear that no such logical development was contemplated.

The unsatisfactory condition of the streets of New York City at the beginning of the present administration is well known. Many of the innovations and improvements introduced under Waring were discontinued and forgotten. Much of the plant was out of order and some of it out of service. Two self-propelled refuse barges were laid up, apparently for no other reason than that they were too economical. Horses were out of condition, overworked, poorly cared for. The dumps and docks were in a disgracefully insanitary condition. The force of laborers was demoralized. The streets were filthy. The rules of the Department, and especially the City ordinances, were shamefully neglected. The policy of entering into contracts with private parties had been extended. The method of letting contracts was open to severest criticism. A contract for the disposal of the garbage of Manhattan had to be let at an exorbitant figure because bids were not called for until too late to secure profitable competition.

Were it not that commerce, convenience and health demand that the collection and disposal of the City wastes be carried on in an efficient manner, the great cost of the Department of Street Cleaning would furnish a sufficient reason why its work should be well done.

The City Budget for 1903 calls for an expenditure of nearly $5 400 000 to pay the running expenses of the Department of Street Cleaning. No other Department of the City Government, except the Departments of Police and Education, spend so much money. The annual cost of the Department of Street Cleaning exceeds the cost of many other Departments combined.

It may be profitable to consider for a moment the chief items of expense, as shown by the Annual Reports of the Commissioners of Street Cleaning:

First: Sweeping.—By this is meant the collection of dirt, etc., from the pavement. This item, in the year 1901, for Manhattan and the Bronx, amounted to $1 402 056.82.

Second: Cartage.—This includes the hauling of refuse material, such as garbage, ashes, paper, and other like refuse. Carting, in 1901, in Manhattan and the Bronx, amounted to $990 577.71. Mr. Soper.

Third: Final Disposition.—This means the reduction of garbage by contract, dumping of ashes and other dry refuse to fill land, and sea dumping. In 1901, final disposition, for Manhattan and the Bronx, cost $733 602.01.

Fourth: The Removal of Snow and Ice from the Streets of Heaviest Traffic.—This item, in 1901, is said to have been about $600 000.

Fifth: Administration.—Administration cost $223 456.62, in Manhattan and the Bronx, in 1901.

Sixth: Rents and Contingencies.—For Manhattan and the Bronx, this item amounted to $121 490.07 in the year 1901.

The total sum represented by the foregoing items is $4 071 183.23.

From the reports of Mayor Grant's Commission, and the Commissioner of Street Cleaning for 1900, the expenses appear to have increased out of all proportion to the quantities of refuse removed. The figures are as follows:

Year.	Appropriation.	Loads of Refuse.
1882	$1 010 000	1 116 360
1890	1 280 525	1 702 102
1900	5 584 100	2 602 120

But the cost of the Street-Cleaning Department, as shown by the official reports, does not represent the total cost to the citizens for removing and disposing of the City's refuse. Other items of expense are the removal of dead animals from the streets, the removal of manure from stables, the removal of market refuse, the removal of cellar dirt, and the sprinkling of streets.

The largest extra is for private carting. An immense number of property holders employ private scavengers. Hotels, office buildings, apartment houses, restaurants, manufacturing establishments, retail dry goods shops, and many residences are compelled to remove their wastes at their own expense. Recently, even the privilege of the City dumps has been denied to cartmen, and it is said that in some cases refuse has had to be hauled several miles and carried across ferries for disposal.

The total expense of the extra scavenging work made necessary by the inability of the Street-Cleaning Department to collect and dispose of our municipal wastes in a satisfactory manner is very large. In 1891 the sum paid for the private collection of refuse was estimated to be between $150 000 and $175 000. It is now very much larger. The quantity of refuse of various kinds dumped at City wharves by private carts has been more than 30% of the total quantity collected by the Department of Street Cleaning. In addition, more than 100 tons of swill per day are carried out of town to be converted into soap, or fed to cattle and swine.

Mr. Soper. In reply to a request from the Board of Aldermen, for a specification or draft of the organization necessary to handle the output of refuse of the large office buildings, stores and places of business which now employ private scavengers, the Street-Cleaning Commissioner said, a few days ago, that he would require 500 additional horses, 500 carts, 500 drivers, 50 hostlers, three stable foremen, three assistant stable foremen, nine stable clerks, twenty stablemen, proper stabling for the horses, and spaces for the carts, together with such equipments as harness, blankets, buckets, etc., and proper water-front facilities for the handling of this material. The total number of carts and trucks now under the Commissioner, in Manhattan and the Bronx, is 744. By making his request for 500 more carts, the Commissioner tacitly admitted that 40% of the work of removing refuse in New York is now done by means outside of his Department and beyond his control.

The private collection of refuse is objectionable, from every point of view. It leads to insanitary practices, and is an item of expense and annoyance from which the taxpayers of the City should be free. The cost of private collection is not equitably distributed, but is often heaviest upon those who pay the largest taxes. But there is no use in emphasizing the evils and extravagances of the present situation. If the force, plant and organization of the Department of Street Cleaning were adequate to the needs of the City there would be no excuse for them.

If the foregoing remarks carry any clear idea, it must be that the work of collecting and disposing of the wastes of the metropolis is badly done. Such efficiency as exists is due largely to the energy of the present Commissioner and the effects of the two notable epochs in the history of the Department to which reference has been made. The first of these epochs was the investigation, report and reorganization due to Mayor Grant's Commission. To this may be added the investigation and the improvements suggested by the Committee of Advice appointed by Mayor Gilroy. The second epoch was marked by the tenure of office of Colonel Waring.

In the history of the Department, few, if any, important reforms or developments have been made which were not the result of seeking outside investigation and advice. Is it not time that the disposition of the City's wastes be again made the subject of an exhaustive inquiry?

At the present moment the City is deplorably dirty. Business men and property holders are complaining that their refuse is not removed, and an immense sum of money is being spent for the collection and removal of wastes which, upon theoretical considerations, are valuable. The Street-Cleaning Department is struggling under the disadvantages of an imcomplete and outgrown equipment. It may seem like strong language, but it is not beyond the facts, to say that the system of cleaning the streets of the metropolis, if followed by a private corporation, would lead to bankruptcy. Mr. Soper.

The speaker intends no unfriendly criticism of Dr. Woodbury, our able Street-Cleaning Commissioner, in stating that the length of his term and the power of his office are not sufficient to enable him to correct, single-handed, the mistakes of the four years of mal-administration which preceded him, and place the Department on the efficient footing which the citizens have a right to expect. To meet the changed conditions due to the recent phenomenal growth of New York City, the Department needs to be developed, expanded and made as efficient as it was in Waring's time.

The speaker sees only one way by which these ends may be accomplished, and that is through the work of a Commission of Investigation and Advice, such as was appointed by Mayor Grant.

In the speaker's view, the most important specific problem pressing for solution is: How shall the City's wastes be finally disposed of? This is the keystone of the great question as to the best means of organizing and equipping the Department of Street Cleaning.

Second.—Another query is: Is it well for the City to enter into contracts with private individuals for horses and carts to collect the refuse, for barges to remove it to a distance and for reduction works to dispose of it?

Third.—One more question is: How can the ordinances relating to the careless littering of streets and the separation of refuse be enforced?

Fourth.—Finally: In what ways and to what extent should the plant of the Department be increased?

Without desiring to amplify this last interrogation unduly, or to presume to possess the knowledge required to answer it, the speaker would state that it is believed, by those who have taken an intelligent interest in the Department, that more and better dumps are needed; more and better scows are needed; more and better carts and horses and stables and stations are needed; and refuse destructors or garbage crematories, built in accordance with the best known practices, are needed. In addition, very possibly, street-sweeping machines and snow-melting machines are needed.

It should not be an open question whether mechanical devices cannot, to some extent at least, replace the hand labor which is now the chief item of expense in cleaning the streets.

The present moment is propitious for an investigation. The existing administration is pledged to municipal reforms, and has executed or planned several admirable lines of improvement. The present Street-Cleaning Commissioner has demonstrated, by a year's administration of his office, that he has the capacity, energy and determina-

Mr. Soper. tion to put in force such measures of relief and permanent improvement as may be placed within his reach. The work of the Commission which the speaker suggests would greatly facilitate his efforts to bring the plant and organization of the Department to a high state of efficiency.

It is a pleasure to say that the plan of appointing a Commission has received Dr. Woodbury's endorsement, and that Dr. Lederle, President of the Board of Health, has given it his approval.

Mr. Horton. THEODORE HORTON, Assoc. M. Am. Soc. C. E.—The Trenton garbage plant was a source of serious trouble during last summer (at a time, in general, when such plants give evidence of any defects in design or management), and a consulting engineer was called upon to investigate and make a report to the City in regard to it. He was asked, first, to determine the nature and extent of the trouble, and, secondly, to make such recommendations as might be needed to relieve the situation.

The details of the work, in a great measure, devolved upon the speaker, and he will describe briefly what occurred during his visit.

The plant, which, unfortunately, was located in the midst of a thickly settled district, consisted of two Davis destructors, connected with a stack 120 ft. high. Each furnace consisted of a combustion chamber; a drying chamber, below which was located an evaporating pan; secondary fire grates; a flue of somewhat contracted area leading to the chimney; and a set of screens to intercept dust particles.

The speaker spent the first day in familiarizing himself with the construction and operation of the plant, and in getting under headway a systematic record of the various features which would lead to a solution of the trouble. These consisted of a record of weights of garbage consumed, coal burned, and ashes; the times of delivery of garbage, charging of furnaces, firing, cleaning grates, operating dampers, and such other observations as might throw light upon the causes of the trouble.

While these records were being kept by the various employees of the plant, other observations, of a special nature, were carried on under the speaker's personal supervision. Thus, he made several analyses of the garbage, which in this particular case was a very important element, due to the fact that it was that known as kitchen garbage, or swill, and was free from ashes. At that season of the year, it was practically saturated. These analyses were simple, and consisted of merely assorting samples, of about ½ ton each, into various parts and weighing them. The garbage was found to consist of 84% animal and vegetable matter, 12% paper, 3% rags, carpets, etc., and 1% boxes, barrels and other rubbish; 80% of the whole being water.

Other observations were made to determine the extent of the

Mr. Horton.

trouble arising from the dust particles which, emanating from the stack, were causes of much annoyance to the residents of the neighborhood, and of alleged fires in the immediate vicinity of the crematory.

The observations made during the daytime were simple in nature, but served at least to give a tangible idea of the actual conditions. They consisted of observing, at various distances from the crematory, and in the line of smoke from the stack, the number of particles passing a vertical plane in a given time. The distances selected were $\frac{1}{4}$, $\frac{1}{3}$ and $\frac{1}{2}$ mile, and the interval of time 15 minutes. The particles were from $\frac{1}{8}$ to 1 in. equivalent diameter, and of course were plainly visible.

In order to present this source of trouble in a still different light, other observations were made by selecting places in the open fields, at various distances from the crematory, and counting the number of particles deposited on an area of 50 sq. ft. During both the foregoing sets of observations, which covered a period of about 3 hours, the wind had been blowing steadily from the same quarter, so that, considered as a whole, the record really had a comparative value, notwithstanding its approximate nature.

In order to get some idea of the danger caused by these dust particles as regards the alleged fires in the vicinity, observations were made upon the glowing embers emitted from the stack at night. Thus, it was observed that:

First.—No embers were seen in a state of actual flaming.

Second.—Glowing embers were emitted principally at times when the furnace doors were open for firing, raking or cleaning.

Third.—During these times it was estimated that, on an average, 50 to 100 sparks were emitted per minute. Of these, it was estimated that 50% did not travel, in a glowing state, further than 25 ft. from the top of the stack; 90% not more than 50 ft.; 95% not more than 75 ft.; 99% not more than 100 ft. and an occasional spark from 150 to 200 ft.

In studying the question of internal combustion in the furnaces, other observations were made, to supplement the general record kept by the employees. These, however, were mostly of a superficial nature and hardly admitted of any refinement, except, perhaps, a determination of the temperature of the flue gases, and even this was somewhat approximate. The speaker had suspected that the temperatures were very low, and at first had tried 500° mercurial thermometers, but, after breaking two of these, and not being able to obtain a pyrometer, he resorted to the somewhat crude method of exposing in the flues metals having different melting points. By a system of elimination, these experiments indicated a temperature of between 800 and 900°, which was sufficiently close for the desired purpose.

Mr. Horton. At the end of the week's run, the foregoing records were tabulated and studied, and the following is a summary of the results:

RESULTS OF OPERATION OF GARBAGE CREMATORY AT TRENTON, N. J., FOR THE WEEK OF AUGUST 4TH–9TH, INCLUSIVE, 1902.

Total garbage burned	188.0 tons.
Garbage burned per day	31.3 "
Coal used on main fires	10.7 "
Coal used on auxiliary fires (estimated)	3.0 "
Total coal used	13.7 "
Coal used per day	2.3 "
Garbage burned per ton of coal	13.8 "
Approximate average time of burning each day	14.0 hours.
Equivalent number of days of 24 hours	3.5 days.
Equivalent garbage burned per square foot of grate area, per day of 24 hours	1 080 lbs.
Equivalent amount of garbage burned per cell of 25 sq. ft., per day of 24 hours	13.5 tons.
Estimated total amount of clinkers from garbage and coal grates	14.0 "
Estimated amount of ashes from below grates	3.0 "
Estimated total amount of ashes and clinkers	17.0 "
Percentage of ashes and clinkers to garbage burned	9.1 per cent.
Range of temperature of flue gases in stack	600°–1 000° Fahr.
Percentage of moisture in garbage	81.0 per cent.
Corresponding water evaporated daily in furnaces	25.5 tons.
Quantity of coal required per day to evaporate this water, on a basis of 10 lbs. of water per pound of coal	2.5 "

A discussion of all the questions involved in the investigation of this plant has been embodied in the report of the consulting engineer, and as this has already been published elsewhere in full, the speaker will give merely a brief abstract of the conclusions reached and the recommendations made to the City, as follows:

"1.—Notwithstanding the existing defects and the criticisms which it has been necessary to make, the Trenton furnaces belong to the better class of such plants in our country, and the troubles and nuisances with which you are now contending, are also more or less common elsewhere.

"2.—The collection in well-constructed metal carts, such as you have, and also the frequency of your collections, are quite satisfactory. It may be suggested, however, that provisions for thorough cleaning of the carts after dumping would be a still further improvement.

"3.—To obviate the nuisances of dust escaping at the stack, either from well-burned or unburned garbage, it is necessary, after securing perfect combustion, to introduce a dust-settling chamber of proper size between the furnaces and the stack, containing baffles and properly placed inlets and outlets. The screens now in the chimney, although useful as a safeguard against accidental emissions of light but large matter, are insufficient to retain all of the matter now escaping from the furnaces. Mr. Horton.

"4.—To reduce further the possibility of unburned particles escaping through the stack, it is important to place dampers in the flues leading from each furnace, as well as in the chimney, so as to afford control of the drafts at all points where and when it is needed.

"5.—The garbage now delivered, when properly drained and carefully burned in a well-designed furnace, is readily combustible, and should not require the addition of as much coal as now used; and at some seasons, except to start fires, and, otherwise, on rare occasions, perhaps no coal may be required.

"6.—We find that the furnaces are not sufficient in size to effect a complete destruction of the material during the hours of the day when the crematory is operated. Continuous firing for 24 hours can relieve this sufficiently. To this end it would be necessary to store the material for a part of the day, and this could not be objectionable in any way.

"7.—To secure a complete combustion of the garbage, more economically than at present, the design of the furnaces should be somewhat altered. The horizontal grates are not the most suitable for the material, which must first be dried, and then with difficulty pulled from one position to another for burning. A sloping grate for burning, and a sloping floor for the drying chamber, from which the material can be gradually and almost automatically moved from the feed-hole to the hottest portion of the fire, permit of a better utilization of the heat, and reduce the difficulties of stoking.

"These recommendations, supplemented by a systematic attention and skillful practice on the part of the Superintendent and his subordinates, would enable them to operate the crematory so that nuisances from odors and dust would cease."

W. J. Baldwin, M. Am. Soc. C. E.—The speaker is wholly unprepared to talk on the subject of garbage cremation, for the reason that such cremation as he has been doing has been in another direction, *viz.*, either hospital refuse destruction, or waste fuel utilization. Mr. Baldwin.

Mr. Soper has suggested the matter of snow melting. There is a great deal of fun in the subject of snow melting, and, were it not that reference to it had been made, the speaker might refrain from alluding to it. The subject is very old, and should be well understood by engineers; still, it comes up, as in the present instance.

When Inspector Williams was made Commissioner of Street Cleaning of New York City, the *Sun* asked why he could not melt the snow off the streets over night, and not leave it to be shovelled and carted when the contractor got ready; and suggested the use of the fire engines of New York for this purpose. Of course, a technical paper would hardly propose such a thing, but Inspector Williams was exercised so much that he asked the speaker if there was anything in the suggestion. The speaker went to Professor Draper, of the Mete-

Mr. Baldwin. orological Department, and ascertained the amount of the snowfall for a certain storm in New York City, and then calculated the street areas, not including the housetops or back yards. A computation showed that it would take all the fire engines in New York City, running day and night, three years to burn coal sufficient to melt that snowfall; that there would be 90 000 tons of coal to haul, and 20 000 loads of ashes left to cart away. The *Sun* would not publish this information, but it appeared in *The American Machinist.* Later, an experiment was made by Mr. Waring, in melting snow, in a manhole in the street, with a coil of pipe and a boiler, which showed the futility of such methods.

Experiences similar to these—steam-heated snow-ploughs, etc.—should set at rest the question of melting the snow, especially with steam, or even with direct fire. The computations were made on the supposition that all the steam generated could be brought in contact with the snow; but, as a matter of fact, in trying to melt snow by any kind of pans, or by blowing steam through it, probably not 10% of the steam would come in contact with the snow, and, in consequence, it would blow away. Again, evaporation will take heat not required for melting. It requires 142 B. T. U. to melt 1 lb. of snow from snow at 32° to water at 32°; but it requires more heat when the snow has a temperature of zero, additional heat to warm the water so that it will not freeze in running off, and heat—a great deal of heat—to vaporize the proportion of the water that will be vaporized.

In regard to particles, or sparks, that appear at the head of a chimney of a crematory, especially at night, they always appear as though they were luminous from heat, but investigation has proved that they were not always so; that immediately over the chimney these particles reflected the light of the chimney, and even when they passed a long distance—probably 20 ft. outside of the axis of the chimney—they would look as if they were afire; whereas, as a matter of fact, it was only the reflected light from the chimney. In an investigation leading to a determination of how far the sparks might fly, it was found that they were not actually sparks at all.

Again, there is the question of what might be done in the way of quenching sparks, assuming that they are sparks. With a chimney 100 ft. high, and having 1 000° of heat, there is a considerable upward push, which might be represented by 2 or 3 ins. of water pressure, and it is possible that something might be arranged at the top of such a chimney by which the sparks, smoke and gases could be given a slight dip into a water tank and thereby cooled.

Mr. Parsons has said that he thought this crematory or destructor could be made to operate without extra fuel. The speaker does not think this would be possible, as there is more than 80% of water in the vegetable and animal matter in garbage. A cabbage will contain more; and all ordinary albuminous compounds—meats and other refuse—contain so much water that the speaker cannot see how it is possible to do without the use of some fuel. Mr. Hering says that the water is evaporated first, or that the moisture is drained off in channels before the garbage is put upon the fire, and is then mixed with a large quantity of dry refuse—wood, straw, paper, etc.—and also a large quantity of unburned coal. Mr. Baldwin.

In that case, the solution seems to be that there is enough carbon present to evaporate the water, but still, in many cases, there is a great deal of water after the ordinary draining and drying is done. An ordinary dry cabbage contains 80% of water, and the water is very hard to destroy, but, of course, it is necessary to take all the material that comes from hotels and from kitchens, as that is what is to be destroyed. Now, the speaker is satisfied that what is collected in separate receptacles as garbage, in New York City, would contain 80% of water which would not run out by gravitation, and that about 10% of the original would have to be good carbon in order to evaporate the contained water.

The speaker would like to see an apparatus which will burn the ordinary garbage and admixture with dry refuse, or even with ashes. The ashes from an ordinary engineer's department amount to about 20%, by weight, of the coal put in the furnace, but in those ashes there is not 2% of coal. Now, there may be wood enough, in some cases, but, even in New York City, where will the necessary quantity come from?

The speaker can show drawings and furnaces constructed for special work—not primarily for destroying garbage, except in a general way—but is interested in the construction of crematories, and would like to know more about the way to construct a furnace and get along without a primary fire and the addition of fuel.

Another point about cremating garbage, and the utilization of the heat therefrom under boilers, is the formation of various vapors, or gases, which are partially condensable when they strike the heating surface of a boiler. They may contain acids which will injure the boiler tubes to a considerable extent; and, in Europe, where they utilize the heat under boilers, presumably, the boilers are of a very simple type. It would be well for anyone who desires to utilize the heat from garbage and general refuse crematories to study the effect of such gases on the small tube boilers used so extensively in America.

Charles B. Ball, M. Am. Soc. C. E.—There are two inquiries which seem to be pertinent to this subject. One is the question as to how far it is safe and desirable to use street sweepings as material for filling. The second relates to precautions to be observed in handling builders' refuse resulting from the demolition of dwellings which have been for a long time in an infected condition. Mr. Ball.

Mr. Ball.

THE USE OF STREET SWEEPINGS FOR FILLING.

The division of municipal refuse into four classes, as used by Mr. Hering:

I.—Ashes;
II.—Street sweepings;
III.—Dry rubbish;
IV.—Kitchen garbage;

is a convenient one for the present purpose. It should be remembered that only in exceptional cases are street sweepings collected separately, the usual method being to mix them, as found convenient, with any other class of refuse.

Premises.—To make clear the proposition advanced by the speaker, let it be assumed, at the outset, that kitchen garbage alone ought not to be used as material for filling low grounds, and that all mixtures of garbage with either ashes, street sweepings or dry rubbish are objectionable for this use to the degree in which they contain the more putrescible ingredients; also, that under some conditions ashes, and sometimes dry rubbish, may be utilized in this manner without danger.

The inquiry, then, is thus formulated: Whether street sweepings alone, or mixtures with ashes or dry rubbish, into which they enter as a principal element, are not undesirable materials for filling in low lands, parks and streets? It seems that this use of sweepings ought to be questioned, for three reasons: First, because of the considerable percentage of organic matter which they contain; second, because of the disease germs which are present in them; third, because of their objectionable physical qualities.

Prevalence of this Method of Disposition.—In connection with the general statements in Mr. Hering's opening discussion, that the use of sweepings for filling low land is a common one, certain data in *The Engineering Record* of November 29th, 1890, are of value. In a tabulated statement giving returns from seventy-eight cities and towns, as to the method of disposition of street dirt—which, doubtless, includes rubbish as well as sweepings—it is stated that fifty-four of these municipalities (69%) used such materials "for filling low places and private lands, or for grading streets and park grounds."

From the more comprehensive statement summarized by A. Prescott Folwell,* M. Am. Soc. C. E., from the detailed reports obtained for the "Municipal Year Book for 1902," it is learned that 741 of the cities and towns having a population in excess of 3 000 report that street sweepings are used for filling. This number is 76% of those that report upon the method of disposal. It may also be noted that the Department of Street Cleaning of the City of New York advertises that:

"Persons having lands or places in the vicinity of New York Bay to fill in, can procure, free of charge, material for that purpose—ashes, street sweepings, etc.—collected by the Department of Street Cleaning."

* *Engineering News*, November 20th, 1902.

Quantity.—Although few statistics are obtainable as to the total quantity of street sweepings handled by municipal authorities, the ratio which the refuse thus classified bears to the total quantity collected is ascertainable for some of the larger cities, as well as the quantities of sweepings collected by them. Mr. Ball.

In a table prepared for the Advisory Commission to Mayor Hugh J. Grant, and published in its report in 1891, the percentage distribution of the refuse collected in New York City in the years 1882–1890, is stated to be:

Ashes	about 62 per cent.
Garbage	12 "
Ice and snow	2 "
Street sweepings	24 "

The statement made by Mr. Parsons is that the distribution by volume is:

Ashes	57 per cent.
Rubbish	25 "
Garbage	18 "

There appears to be no difficulty in harmonizing these statements, it being understood that street sweepings and rubbish are included in a single item, and that the garbage constituent may often include many kinds of waste which, in a strict analysis, should be classified as rubbish rather than garbage. The latter table, by Mr. Parsons, showing the *per capita* distribution of the refuse in Greater New York, gives about 18% by weight of street sweepings. From these, and other less complete data, it would seem probable that 20% of the total amount of refuse represents well the proportion of such material in cities where the collection is made in well-defined classes, or that, at all events, such a percentage is not too large.

The report of the operations of the Department of Street Cleaning for the year 1900, which is the latest obtainable, states that it collected from all of Greater New York, 2 307 594 cartloads of "ashes, street sweepings and other refuse," apart from the item of garbage, separately named.

Composition of Sweepings.—The composition of street sweepings varies greatly with the character of the pavements from which they are gathered, with the method of sweeping—whether by hand or machine—and with the effectiveness of the separate collections of dry rubbish, such as dust, leaves, rags, paper, metal, old leather, straw, wood, glass, etc.

Organic Constituents.—It is obvious that sweepings from smooth, impervious, elastic pavements will contain much greater percentages of organic matter than such as are collected from macadam or rough stone pavements.

The limits indicated by Baumeister* are 4% for macadam pavements to 23% for stone pavements, and 60% for wooden pavements.

* "Cleaning and Sewerage of Cities," 1891 edition, p. 177.

Mr. Ball. An analysis of street dirt made in 1874 by Professor Edwin Waller, for the Health Department of the City of New York, gave, for an average of thirteen samples:

Water	3.032	per cent.
Organic and volatile matters	28.454	"
Nitrogen	0.369	"
Mineral	68.514	"

There appears to be no evidence to determine whether rubbish was or was not included in the samples for this analysis.

Examinations of the percentage of moisture, ash and organic and volatile constituents found in samples of ashes, sewer dirt, street sweepings and rubbish collected in New York City on two different dates, March 14th and May 1st, 1902, give percentages of the organic constituents varying from 17.86 to 40.70. The average of five analyses for each of the above dates was 26.31% of organic and volatile matters. The samples were collected by the skilled chemist who made the analysis, and garbage was excluded.

The prevalence, at the present date, of sheet-asphalt for street surfaces, as well as the constant increase in the ratio of paved to unpaved streets, would be likely to indicate that a higher percentage of organic matter is common, and it would seem that an estimated percentage of organic matter in sweepings at least as high as 35% would be justified.

Rubbish Constituents.—If street-cleaning machines are used instead of hand-sweeping methods, the resulting refuse will contain a smaller quantity of miscellaneous rubbish, as the sweepers and shovelers in the latter case collect from the gutters various articles of small dry rubbish which the machines do not touch. Most of these materials, such as leaves, paper, straw and hay, are light and combustible, but considerable quantities of waste metals, glass, crockery, plaster, etc., are heavy and incombustible.

Germ Constituents.—In addition to the tuberculosis germs noted by Mr. Hering as being abundant in street dirt, the bacilli of diphtheria are by no means uncommon, and typhoid fever germs are sometimes found.

A series of experiments was conducted by the Street-Cleaning Department of the City of New York in August, 1902, by exposing Agar plates, at the curb level and 6 ft. above, in various locations distributed over the city, and noting the number of colonies of bacilli developed during a stated period. The isolation of specific pathogenic germs was not attempted, to any considerable extent, the object being to determine the varied effect of different kinds of traffic in distributing the bacteria in air; but a maximum of 30 colonies per inch-minute was found in the east-side tenement district, and a minimum of 0.06 colony for a suburban street in the Bronx.

The important question respecting the danger from these bacilli is that of their viability. Mr. Ball.

It is now well known that the tubercule bacillus retains its life properties for long periods when dry, but it is not so well demonstrated that it is dangerous when in a moist condition, such as would occur most frequently beneath the surface of a fill. Respecting the life of the typhoid and diphtheria bacilli, there appears to be no doubt that these are perpetuated in damp surroundings.

Professor Charles Harrington, in his "Manual of Practical Hygiene" (1901), discussing the function of the air as a carrier of infection (pages 226–230), summarizes the extensive researches of Dr. Eduardo Germano on the permanence of typhoid with the statement that "this germ as well as that of diphtheria can live for a long time in moist surroundings."

G. Lister Sutcliffe, in an introductory chapter in his "Sanitary Fittings and Plumbing" (1901), says:

"Dr. Sidney Martin's experiments with the typhoid bacillus, for example, prove that it soon dies in 'virgin' soils, whether these are sandy or peaty, but that it rapidly multiplies in moist sterilized samples of soils from gardens and other cultivated places, and retains its viability and vegetative properties for months and perhaps years. Moist sterilized soil which had been impregnated with the typhoid bacillus was found to contain the bacillus at the end of 450 days. The soil was then naturally dried for 49 days, till it 'could be readily powdered into a fine dust;' the bacillus was still there.

* * * * * * * *

"The lessons to be learnt from these and other investigations are that some pathogenic organisms are not by any means so short lived as is commonly supposed, that they may spread rapidly under suitable conditions, and that they may exist for a considerable time even in dust so dry as to be blown about by the wind."

Physical Characteristics.—It is evident that the physical qualities of permanence and stability, required in ground to be used for foundations of structures, or even for the sub-surface of streets and parks, are lacking in street sweepings, owing to the considerable quantity of organic matter and light combustible material contained.

Conclusions.—That the opinions expressed by the speaker are not altogether new, is shown by a paper on "City Scavengering at Boston," read by Eliot C. Clarke, M. Am. Soc. C. E., at the meeting of the Metropolitan Public Health Association in Nashville, Tenn., on November 18th, 1879. The work of scavengering is here classified as constituting four parts:

1.—Removal of offal;

2.—Removal of ashes and house dirt;

3.—Cleaning of streets and street basins;

4.—Cleaning of privy vaults and cesspools.

Under the second division of his subject, Mr. Clarke writes:

"A large part of Boston being very low land, there is great temptation to use this refuse (ashes and house dirt), in the place of earth filling,

Mr. Ball. and such is its usual disposition. If the ashes were perfectly free from admixture with other substances, it would form suitable filling for most purposes, but it is not so."

He then explains the attempts to keep out objectionable rubbish, and, in the following statement, summarizes the failure to accomplish this:

"On an average, nearly one-third of each cart-load of so-called ashes consists of other refuse. Portions of this refuse are harmless enough, as, for instance, broken glass, crockery, plaster, old iron and tinware, but other portions entirely destroy the suitability for filling, such as bones, hair, leather, rubber, paper, pieces of cloth and carpeting, old baskets, and, in short, any vegetable or animal matters liable to decompose. The ashes and house dirt are used to grade streets and parks belonging to the city, and private lands, when desired or permitted by the owner. In a few cases a small sum is paid to the city for the filling. It is evident that the presence of the refuse house dirt renders ashes undesirable for street building, as the matter, as a whole, compacts very slowly, and continues to settle for years. If the surface be paved or covered with gravel, there will probably be no smell; but digging trenches for laying gas and water pipes, sewers and drains, will be an offensive operation for a number of years, since organic matters, kept from a free supply of oxygen, decompose slowly.

* * * * * * * *

"It is probably a serious mistake to dump house dirt on land which is to be built upon. It is true that the organic matter mixed with the ashes must in time disappear; but how soon, it is impossible to say. Excavations made in Boston in land filled from city carts ten years previous have disclosed remnants of organic matter, and the process of decomposition is still going on. The risk in living over such land may be slight, but, however slight, it is a risk of disease and death."

In considering the third division of the subject, he writes:

"Forty-eight thousand and fifty-nine (48 059) cart-loads of dirt (street sweepings) of 40 cu. ft. each, were collected during the past year (1878?), and were disposed of in the same manner as ashes and house dirt, by being used to fill land for streets, parks and dwelling lots. The street dirt is of a rather uniform character, consisting of earth, road detritus, fragments of wood, leaves, paper, soot, hair, etc. Under exceptional conditions, from well-paved streets in the heart of the city, a load is sometimes obtained of sufficient manurial value to fetch a small price for use on land, but this is seldom. It is a manifestly unfit substance for filling, both from an engineering and from a sanitary point of view; but what else to do with it is a problem yet unsolved."

To the speaker, the solution appears to be a clear one, in view of the statement by Mr. Parsons, that "rubbish collections are very combustible, as are also street sweepings." The conclusion appears to be incontestable that street sweepings may and should be burned, irrespective of the question whether they can or can not be made to assist in disposing of the more objectionable garbage.

THE DEMOLITION OF INFECTED DWELLINGS. Mr. Ball.

It has been recognized for some years by students of sanitary science that human habitations in which filthy conditions prevail for a long time become so infected with disease germs that no cure can be effected. Although the recent methods of disinfection by formaldehyde gas are to a degree effective, it seems questionable whether even those will render safe for habitation a dwelling which has for a long time been permeated with disease and filth.

Danger of Infection.—The report of the Tenement House Committee of 1894 (usually called the Gilder Commission, from R. W. Gilder, its Chairman), makes a brief reference to their conclusions in this regard, as follows:

"It is a well-known fact that human dwellings, by prolonged occupation, become so saturated with the bodily emanations of those who have lived in them, sick and well, as to be less healthy than newer houses. It has often been proposed to construct hospitals of flimsy and inexpensive material, after the manner of a temporary shelter, and destroy them by fire every few years, to be replaced with new ones.

"This result of prolonged human occupancy is made worse in this city by the rapidity with which the character of the population in the lower parts of the city has changed, and the fact that houses built as private dwellings are not fitted to serve as tenements, either in respect to light, ventilation, water supply, privacy or facilities for cleanliness."

Respecting the fact of tuberculosis infection in old houses, there is no higher authority than Arthur R. Guerard, M. D., of New York City. In a paper presented to the Tenement House Commission, November 16th, 1900, treating of "The Relation of Tuberculosis to the Tenement House Problem," he says:

"It has been repeatedly observed that some houses in overcrowded districts after a time seem to become permanently infected with tuberculosis, as evidenced by a repetition of cases of the disease. * * * It is easy to understand how the permanent infection of premises may occur when one considers that the sputum of consumptives, which frequently contains a countless number of virulent tubercle bacilli, is often carelessly expectorated on the floor and elsewhere in dwellings, where it dries and is converted into dust, thus getting into cracks in the woodwork, in the papering, etc., whence it is later inhaled into the lungs by others. Such tuberculous dust has been found by experiment to remain virulent and infectious for months and even years after it was deposited in rooms which were not thoroughly disinfected and renovated; and where this dust is allowed to accumulate for a length of time it may be impossible to disinfect and renovate the premises without practically rebuilding them. These premises then become permanently infected."

John F. J. Sykes, M. D., in his recent book on Public Health and Housing, thus refers (pages 82 and 83) to the dangers of infection:

"Miquel, Emmerich, Carnelly, Haldane, Anderson, Johnstone and other observers and experimenters have shown that with the in-

Mr. Ball. creasing use and age of a dwelling-house the walls and spaces beneath floors become liable to harbour micro-organisms to an increasing degree. The more the surfaces, especially the floors and walls of a house, become worn and defective, the better lodgment they present for the accumulation of organic matter—the nidus of micro-organisms—and the less attention paid to repair and redecoration, the more unhealthy the influence of the house becomes.

"All these experiments and observations indicate the necessity for such construction, maintenance, and usage as best conduce to the cleanliness and purity of the air in and around dwelling-houses, in order not only to prevent disease, but also to prevent the loss of that vitality, vigour, and energy which cannot easily be measured by experiments or by statistics."

The conclusions of these various investigators as to the danger in the use of such dwellings, appear to the speaker to justify reasonable precautions in their demolition, and in the handling of the materials, both the valuable and the useless, resulting from this tearing down.

The number of buildings for which demolition permits were issued in 1902 by the Bureau of Buildings, Borough of Manhattan, Greater New York, was 1 622. This number, it is well understood, does not represent all the buildings demolished, as a small proportion is torn down without the formality of a permit. If this list is examined by anyone conversant with the city, he will note, from the street locations, that certain of these demolished dwellings are very old buildings in tenement districts, and in some cases such as were not constructed originally for tenement purposes. A search of the records of these buildings will show that they are infected with disease, and have been for a long period in unclean condition.

Present Method of Demolition.—The present method of demolition results in the removal, for fuel purposes, by the poor of the neighborhood, of the smaller portions of the wooden floors, casings, roof, etc., which are then stored in the adjoining cellars. It results from this that a building infected with disease germs, filth and vermin, may readily disperse these dangerous elements in the vicinity.

The only supervision of a sanitary nature over the practice of demolition in New York City at the present time is the requirement by the Department of Health that frequent sprinkling with water shall be given to the walls as they are torn down, in order to avoid objectionable dust.

Desirable Precautions.—It would seem to the speaker desirable to examine all buildings, about to be demolished, as to their sanitary condition, and, if this is found to be at all doubtful, to observe the following precautions in their destruction: First, that the building should be kept constantly wet with a disinfecting solution as the work of tearing down proceeds. From what has been stated already, it is evident that the old paper on walls and ceilings, the plaster of walls, and the woodwork of the floors, doors and trim of the rooms and halls are the elements of danger. These should be sprayed thoroughly at the commencement of the work, and kept wet. The Building Code of the City of New York requires the demolition of buildings to be carried on story by story, and the systematic execution of the work as indicated herein would be facilitated by the adoption of this rule in all cases. Mr. Ball.

Second, that no combustible material should be distributed in the neighborhood, but that it should all be removed, and that which is valuable for fuel only thus used under conditions which would prevent the danger of contagion. The old paper, lath, floors, doors and trim of the rooms should be burned in some locality remote from any residence, and, by the use of simple gratings, will furnish fuel for the incineration of much of the light refuse.

Third, that any portions of the structure which are suitable for use in new buildings should not be used thus without having been subjected to suitable disinfecting processes.

Fourth, that, during the process of further excavation of the site, the soil removed should be examined as to polluting matters, and, if found to contain such, it should be treated with disinfectants. The actual demolition of old tenements in New York City, with the excavations for cellars of new buildings, shows that, where school sinks and earthenware house drains have existed, there is an extended saturation of the soil. The removal and proper treatment of this saturated soil is a matter of great importance.

Fifth, that all non-combustible waste refuse should be kept under supervision, as to the method of disposal. Where it is impracticable to treat any waste material by some process of sterilization, the disposal should be at some locality not likely to be used, even in the remote future, for residence or business purposes.

M. N. Baker, Esq.—It is very evident from the discussions that refuse disposal is a distinctively engineering problem; but, thus far, in America, it has rarely been recognized as such. It is to be hoped that the time is coming—in fact, that it has arrived—when there will be a much needed change, in this respect. There are, in this country, connected with the various companies who are building garbage furnaces or who are establishing reduction plants, comparatively few engineers engaged in the work. Mr. Baker.

It is a significant fact that, whereas our municipalities call in eminent engineers and spend money unstintedly for water supply and sewerage, only on one or two occasions have they called in an engineer to advise them in the matter of garbage disposal, and as far as entrusting to an engineer the whole problem of refuse collection and disposal it is doubtful whether a single instance could be named.

There is the valuable report on the Trenton investigation, mentioned by Mr. Horton, but it is yet to be heard that there have been any decided practical results from it. But, at Trenton, no engineer was called upon until after several years of blundering upon the

Mr. Baker. subject; and, bad as Trenton is, as an example of neglect in this respect, it still remains a shining example of almost the only city in the country that has secured any engineering advice whatever.

One of the cities to adopt garbage cremation recently is Milwaukee. The contract for installing the furnaces provided that the contractor should guarantee that they could be operated with eighteen men, but it is stated that there are 79 or 80 on the pay roll at the present time. It is claimed that this is no fault of the furnaces, being a mere matter of affording employment to municipal laborers; but, however that may be, it serves as an illustration of the unsatisfactory condition of garbage disposal in this country, and of the need of employing engineers to design and operate garbage disposal works.

The low temperature in the Trenton furnaces is remarkable. In recent English installations the aim is to get as nearly as possible to a temperature of 1 800 or 2 000° Fahr., instead of 800 to 1 000 degrees.

Some interesting letters have appeared recently in the London *Electrician*—letters brought out by a paper* by W. F. Goodrich, Assoc. Inst. Mech. E., on a combination of refuse destructors (as they call them in England) and electricity works. Several of the writers of these letters, men in some instances who are operating combined refuse destructors and electricity works, have expressed themselves strongly in opposition to such a combination. Certainly, such opposition is a significant fact, and one that should be considered very carefully at the inception of garbage disposal as an engineering problem in this country.

Thus far in the discussion, comparatively little has been said regarding reduction as a means of garbage disposal. To present the subject fairly would take much time, but attention may be called to a fact which, perhaps, has already been mentioned, but only incidentally, that in all, or nearly all, the larger cities of the United States, which have any improved means of garbage disposal whatever, reduction is being used, and scarcely any of them are using cremation.

This does not necessarily imply that reduction in this country has been eminently successful and cremation has been a failure, but the figures do show that reduction is at least worthy of serious attention. All the reduction plants are in the hands of private companies, and comparatively little satisfactory information regarding their financial success is available. Possibly before this discussion is closed it might be feasible to bring out some detailed information from persons familiar with reduction systems. Unquestionably, such information would add very materially to the value of this discussion, in which the many large reduction plants of the country have been virtually ignored.

* "Electricity from Refuse—The Case for the Modern Destructor." A paper read before the Manchester Local Section of the Institution of Electrical Engineers, and published in abstract in *The Electrician* for November 28th, 1902. The letters mentioned appeared in the issues of the same journal for December 5th and 19th, 1902.

Mr. Baker.

CITIES OF THE UNITED STATES, HAVING POPULATIONS OF 200 000 OR MORE, WHERE IMPROVED METHODS OF GARBAGE DISPOSAL ARE USED.

	Population.	Improved means of disposal.
New York	3 437 202	Reduction.
Chicago	1 698 575	None.*
Philadelphia	1 293 697	Reduction.
St. Louis	575 238	Reduction.
Boston	560 892	Reduction.
Baltimore	508 957	None.†
Cleveland	381 768	Reduction.
Buffalo	352 387	Reduction.
San Francisco	342 782	Cremation.
Cincinnati	325 902	Reduction.
Pittsburg	321 615	Reduction.
New Orleans	287 104	None.
Detroit	285 704	Reduction.
Milwaukee	285 315	Cremation.
District of Columbia	278 718	Reduction.
Newark	246 070	None.*
Jersey City	206 433	None.
Louisville	204 731	Cremation.
Minneapolis	202 718	Cremation.

Statistics, in the above table, collected by the speaker for his "Municipal Year Book," and revised to date, show that, of the nineteen cities of the United States having populations of 200 000 inhabitants or more in 1900, fourteen have some improved means of disposal, and, of these fourteen, ten depend on reduction and four on cremation. In addition, two other places have awarded contracts for final disposal, with the understanding that reduction will be used. Even if the list be extended to include places of 100 000 inhabitants or more, having improved means of disposal, there are but six with furnaces to fourteen with reduction plants, or sixteen of the latter if plants under contract are included. Taking into consideration all places of 3 000 population and more, there were, in 1902, a total of 74 crematories and 22 reduction works, all but three of the latter being in places of more than 40 000 population. Reduction is generally held to be impracticable in small cities, but the figures given show that in the United States that system prevails wherever it is feasible.

ELMER W. FIRTH, Jun. Am. Soc. C. E. (by letter).—The use of public dumping places within the city limits is an important phase of refuse disposal as applied to New York City. In the Borough of Brooklyn, for instance, there are deposited at these places, not only Mr. Firth.

* A small quantity of garbage is, or has been, cremated.
† Contract awarded, with the understanding that reduction will be used.

Mr. Firth. ashes, but the dust and dirt gathered from public and private buildings and conveyances, the soiled sawdust from wooden cuspidors, street sweepings, and much garbage and other refuse. Much of this deposit is subsequently blown about the streets in the vicinity, and the Health Department has received many complaints as to the frightening of horses, and other unpleasant experiences resulting from flying papers, etc., from these dumps.

The transportation of such large material suggests a greater distribution of dust and dirt through the air in these localities, and, possibly, an increased morbidity or mortality due to this condition. Figuring the average diphtherial death-rate in Brooklyn, by wards, from statistics in the Health Department reports for 1895 and 1896, it was observed that this death-rate was higher in the vicinity of the public dumping grounds than in other sections of the city. The writer hesitates to mention this as more than a coincidence, admitting that the relative number of cases of sickness would be a better basis for comparison. Then, also, it may be said that as these dumping places are generally located in the poorer sections of the city, scanty or inefficient medical service would increase the number of fatal cases in these localities; but against this argument may be placed municipal supervision, quarantine and the free distribution of anti-toxine.

It should be noted that in diphtheria the seat of infection is in the respiratory passages. The bacillus recognized as its cause withstands prolonged drying, and may be carried through air in a virulent condition. While it is not claimed that this specific disease is transmitted directly by the dust blown about from these deposits of refuse, yet a predisposition to it may result from inhaling constantly such impure air.

Even if the deposition of such refuse is not proved to be accountable directly for much of the suffering in these places, it at least maintains an unsanitary condition in the very communities where the death-rate is highest and in the midst of classes least capable of avoiding infection or of curing disease.

Mr. Falk. MYRON S. FALK, Jun. Am. Soc. C. E. (by letter).—Many and various have been the methods proposed for the sanitary disposal of municipal refuse. The complete destruction of the material, by means of heat generated by itself or by the aid of heat supplied from extraneous sources, has been the method most advocated. The first case brings into consideration the hope that power may be obtained from this waste material at the time of its destruction. In relation to this, and during this discussion, the plan of producing gas, and thus, indirectly, power, immediately from municipal wastes has not been considered; it is possible that mature reflection may reveal great inherent advantages to be derived from this method of disposal.

In the production of gas, as is well known, water in the form of steam is added to the gases produced from the destructive distillation of coal. It seems not at all unreasonable, therefore, in view of the large percentage of moisture contained in garbage, to expect that the introduction of garbage into a gas producer may not only be a direct saving in the fuel required, but may also obviate the necessity of adding live steam. It is true that, in the ordinary case of gas production, it is steam and not wet fuel which is added, but it is not unreasonable to expect that some method can be devised by which garbage may be placed in the producer at such times and in such a manner that its contained water may replace the water which must ordinarily be added. There is but little cause to doubt that such a plan may be put in successful operation, and it certainly seems to offer as great a possibility of obtaining power from wastes as any other plan heretofore proposed. Mr. Falk.

Charles H. Haswell, M. Am. Soc. C. E. (by letter).—The introduction of water into towns and cities for domestic and public services, other than that obtained from public and private wells, involves the laying of pipes for its reception and transmission, and of sewers for its discharge in combination with the excrementitious matter of their residents and animals, the filth of hospitals, households, slaughter-houses, stables, manufactories, tanneries, street washings, etc. The resultant of this flow is the pollution of the earth or water over or into which it is discharged. Mr. Haswell.

The disposal of crude sewage and its treatment, the mitigating of its septic elements, its sanitary, useful and effective disposal, combined with the resulting pollution of the rivers into which it is discharged, and the insanitary effects thereof, is becoming a subject so important, as relates to the existence and duration of human and animal life, that it imperatively demands immediate consideration and action. It is of more national importance than the aggregation of all the financial and political questions of the time; yet, with the exception of the consideration and report of the St. Louis Commission on the Chicago Drainage Canal, and possibly some other isolated cases, it is criminally neglected.

Primary Disposal.—In internal villages or small towns it is customary and quite practical to mix the excreta with earth, bear it beyond the residences of inhabitants, and use it for fertilization, or deposit it in masses and cover it with earth. In the more populous towns of Europe, trenches are excavated and filled, the ground selected being suitable for infiltration of the fluids of the sewage reservoirs and cesspools, and being located at a proper distance from residences. When filled, these trenches are covered with earth, the surface of their contents, meanwhile, being exposed to evaporation, and their whole contents to infiltration (their masonry being open-jointed); but when the earth in which they are placed is clay, wet or saturated, or is con-

Mr. Haswell. tiguous to a stream of potable water, they are constructed with close joints, and, periodically, their contents are pumped out and distributed for fertilization.

Treatment.—In order to mitigate the noxious exhalations of crude sewage, it is, in some instances, delivered into tanks, reservoirs or coke-beds, for aeration, subsidence, filtration and oxidation.

Chemical Purification and Precipitation.—Chemical purification is effected by a strong mixture of lime, salts, sulphur, oxidizing compounds, etc., with the crude sewage. Iron and aluminum salts, if followed by lime, give a purified effluent.* Precipitation is effected by the use of lime in solution, and by sulphate of alumina. In populous towns and cities, any like treatment of their sewage, in consequence of its volume, is wholly impracticable.

Surface Precipitation.—This involves the occupation of a large area of land, it being officially estimated at one acre for every hundred of human inhabitants.

In the course of purification, oxidation is an essential portion of the changes through which sewage has to pass. Sedimentation or subsidence of the solid and dense material of sewage is claimed to arrest pollution; if so, the effect is retroactive, as it would result in the gradual filling up of the bed of the river, by the formation of foul and noxious shoals. The natural purification of water, to the very slight extent to which it is attainable, is effected by the combination of ammonia, arising from the mixture of decomposed animal matter, with the oxygen of the atmosphere.

Filtration.—In order to arrest bacteria and organic matter by this process, sewage is delivered into reservoirs and subjected to the absorbent capacity of coke and sand; which operation involves the frequent cleaning or replacing of these materials, as their absorbent capacity will soon become inoperative on account of the arrest of the dense particles of the sewage.

Infiltration.—Infiltration is classed by Rideal under three divisions as Broad, Intermittent, and Mixed.

Broad.—In broad infiltration, the sewage is distributed over large surfaces of agricultural land, or sewage farms, as they are termed, on which it is purified. This method involves the occupation of very large areas of land, officially estimated, like that of precipitation, at one acre per hundred of population.

Intermittent.—This is effected by the deposit of sewage at short intervals of time, on an area of porous earth as restricted as will absorb it, but making the products of vegetation of secondary importance.

Mixed.—In the mixed system, the sewage is first purified by sedimentation or by chemical means. The efficiency of this operation

* Samuel Rideal, D. Sc., London. John Wiley & Sons, N. Y.

depends wholly upon the character of the soil, as to its porosity and freedom from water, saturation, clay, peat, or conglomerated gravel. The Local Government Board of England prescribes for stiff clay one acre for every twenty-five human inhabitants, and for loamy gravel one acre for every one hundred. Mr. Haswell.

Cremation.—Cremation is effected by the evaporation and the combustion of the crude sewage at a very high temperature, a forced draught in the furnace, and a very high chimney to dispel the noxious effluvia of the combustion. In the operation of some furnaces in England, it was essayed to prove that sludge (precipitated and dense sewage), both remuneratively and without offence, could be incinerated, and that the coke produced paid for the operation. It was found, however, that the agricultural value of the residuum was so much impaired that it could not be disposed of profitably for that purpose. The evaporation of the liquid, and the combustion of the sludge requires the consumption of some fuel. When the sludge is submitted to a filtering pressure, an average of 7.74 tons (2 240 lbs.) per million United States gallons of crude sewage is obtained. At Southampton, England, it was essayed to burn sludge remuneratively, and without noxious results. The combustion of coal was first used to try it, the ammonia evolved was collected and the volatile matter was passed through the fire. It was claimed that the value of the coke produced paid for the cost and expenses, and that the net cost of the combustion and incineration was inconsiderable. In support of this operation, it is to be considered that even where sewage is discharged into bays upon the sea coast and below the confluence of the tidal and river flows, the velocity of the ebb tide is not wholly effective in transporting it to the ocean, as it will subside in slack water, and in the basins (locally and erroneously termed slips) formed by the projection of piers, thus creating noxious exhalations, and causing frequent dredging and the transportation of their contents to the ocean.

Fluid Discharge and River Pollution.—Where towns and cities are located on a long river, and above the extreme line of confluence and commixture of its water with the tidal water of the ocean, as some on the Hudson, Delaware and Connecticut in the East, and on the Ohio, Illinois, Mississippi and Missouri in the West, the pollution of the water by the discharge of crude sewage into it is increased progressively by the discharge from other and adjoining towns, but somewhat mitigated by the gradual addition of water from tributary streams (which, however, will eventually become polluted in like manner), down to the point of its confluence with the water of the ocean.

Effect of Pollution.—It is generally, if not universally, claimed that foul or polluted water purifies itself; and, in support of this claim, tests and observations have been made. Thus, the Massachusetts Board of Health deduced that if the crude sewage was less than 0.0077

Mr. Haswell. ($\frac{1}{130}$) of the water of a river into which it was discharged it would not affect the water objectionably. In other words, it is advanced that if one measure of crude sewage is mixed with 130 measures of river water in its normal condition it would yet be potable, notwithstanding that its septic or putrefying quality would not be negatived or eliminated. Further, an investigator in Europe has declared that if the crude sewage does not exceed 0.066 ($\frac{1}{16}$) or 6.7% of the river water, when it flows with velocity equal to that of the sewage, it would not affect the potable quality of the water; notwithstanding that, in addition thereto, all streams and rivers are polluted more or less by the rain washings from farmyards, roads, decayed vegetable matter and the putrefaction of dead animals.

As to the distance in which crude sewage discharged into a flowing river will become innocuous by the alleged self-purification of the water, authorities differ, as one puts it at 25 miles and another at 200. The desirability of assigning any general distance, assuming it practicable to do so, is negatived by the point of location at which any putrescent matter was present. Thus, a putrefying carcass of an animal, by its gravity (for instance, that of a dog with a weight attached to it), may remain at the bottom of a river. Over this the water would flow, and receive putrescent matter from it for a long time.

While it is conceded that pure water is only chemically obtained, in a laboratory, natural water is acceptable for domestic requirements, provided it is free from noxious matter, and is not offensive to taste or injurious to health and commercial requirements.

Notwithstanding the popular declaration that even crude sewage, discharged in flowing water, will, when mixed with it, in various assumed and empirical proportions and assigned distances of flow, become potable, a knowledge of the various impurities of such polluted water is not pleasing to reflection. The riparian inhabitants of the river into which the sewage of a populous city flows are passively subjected to the drinking of its water of pollution, yet, at the same time, they will consider the presence of any, even slight, impurity in their other fluids or foods, as milk, mineral waters, flour, meats, fruits, etc., as objectionable and not to be submitted to passively. As to the "self-purification" of water, it is difficult to manifest how the many and various chemical, vegetable and animal pollutions can be eliminated from polluted water by any process, physical or chemical, and unless they are eliminated, the assertion that water will purify itself is a common and grievous error, and, despite the claims of chemists and scientists of all grades, they are challenged to manifest in what manner, chemical or self-acting, water infused with excrementitious matter, the discharge of tanneries and slaughter-houses, paint, ink and glue factories, hospitals of all kinds, dead bodies of men, animals and fish, can be made acceptably palatable. A whole is an aggregation of its parts, and if a volume of foreign or offensive matter, however small, is introduced into pure water, its purity is negatived. Mr. Haswell.

Insanitary Effects.—Living animals cannot retain health while the discharged products of their vital action are permitted to be retained about them, and when they are confined in that condition death gradually ensues. The result of a sanitary disposal of fluids is illustrated in the improvement of the health of towns by the reduction of their death rate. An instance is the City of Leeds, England, where, in twenty years, the death rate was reduced from 28 to 18 per thousand of its inhabitants.

In towns and cities where infiltration is wholly depended upon for the disposal of sewage, the soil becomes contaminated with fæcal matter, which, undergoing but little oxidation and in combination with the periodical cleaning out of the cess-pools or reservoirs, renders the exhalations therefrom both noxiously offensive and insanitary. In the country and in some small towns it is customary to avail of streams over which to set privies, and when the streams are sluggish the fæcal matter will subside in its flow and pollute the earth to an extent affecting its sanitary condition.

Reviewing, then, the elements submitted, it is presented:

1st.—That for internal villages and small towns, sewage can be readily disposed of by infiltration in and the fertilization of the adjoining soil.

2d.—That for similarly situated and larger towns and small cities, while the sewage can be infiltrated to a serviceable extent and the solid portion used for fertilization, yet the area of soil required for effectual operation is so great as to render this method generally, if not wholly, impracticable.

3d.—That from all locations, from villages to populous cities, on tidal rivers or sea coasts, up to the clearly defined limits of the tidal flow or confluènce with the fresh-water flow, their sewage may be discharged directly in its crude condition, as instanced in the seabound cities of Boston, New York, etc., Burlington on the Delaware, etc.

4th.—That if the discharge of sewage, crude or treated, into our rivers is continued, it will wholly arrest them from being resorted to for the supply of potable water, and supplies must be obtained from the ponds and lakes with which our country abounds. Animal life cannot exist in polluted water, and the cost of arresting its pollution, in view of its result, is not to be considered. As the health of our people is of primary importance, the emergency must be faced.

5th.—That to draw upon the waters of Lake Michigan and its northern tributaries for the service of the Chicago Drainage Canal and the mitigation of its pollution of the rivers into which it is discharged, and by assigning a given volume of it, *pro rata* with the number of popu-

Mr. Haswell. lation of the City of Chicago, would result in the lowering of the surface levels, and all adjoining waters down to the Gulf of St. Lawrence, and be prejudicial to both the manufacturing and commercial districts of all the intervening towns, as a consequence of which it would not be permitted, either by the Government of the United States or that of Canada. Hence the proposal is utterly impracticable of adoption, as its effects would soon become so insupportable to the people of both countries that they would arise in opposition, and Congress would be appealed to for relief, and damages be claimed.

6th.—As the sewage of populous towns and cities cannot be disposed of effectually, or with sanitary requirements, either by evaporation, precipitation, filtration, or its discharge into rivers above the line of commixture of tidal and fresh waters, or into lakes, or by mixture with great proportional volumes of water from our lakes, with immunity from the pollution of the potable waters thereof, there are but two methods left, viz., one for populous inland towns and cities, the sewage of which must be chemically discharged of its noxious exhalations and disposed of for fertilization, if practicable to do so; and the other by cremation, in which there is no danger of the pollution of water supplies, or insanitary effects. To operate it, involves the construction and maintenance of a furnace plant, and that the sewage, as received, must be as free from surface fluids as practicable, and that where sewers for the reception and conveyance of both sewage and surface water exist, as in large cities, independent drains be laid for the surface water.

7th.—That, great as this combined cost of reconstruction and construction would be, it is so inconsiderable, compared with the beneficial results to be attained by it, that the expenditure is not to be entertained as a bar to the method, for the result of the discharge of the water of our rivers above the line of commixture given will soon have to be faced squarely, as the destruction of the potability of the water of all rivers bordered by populous towns or cities will become eventually so extensive that it will interfere with the requirements of the people and of their horses and cattle. It will also cause the destruction of the fish, and how it is that the citizens of Chicago entertain the opinion that the operation of their drainage canal will be submitted to, and that the riparian residents of the rivers into which its sewage is discharged will submit to it passively, is wholly inexplicable to the writer.

8th.—That the discharge of sewage, crude or treated in any antiseptic manner, in a fresh-water river, is offensive, noxious and not to be entertained with any regard whatever to the sanitary requirements of its riparian residents; that it will not be submitted to; and any or all towns or cities contributing to the pollution of such water will not only not be permitted to continue in such practice, but will inexorably be required to dispose of their sewage in some other and proper manner. Mr. Haswell.

9th.—That existing wells, for households, in densely populated towns and cities should be filled up and abandoned forthwith, and the vaults of all privies and tanning vats made water-tight, to arrest infiltration of any part of their contents to contiguous and subterranean watercourses, and to guard against the result of the ignorance and indifference of some people who disregard all sanitary requirements and permit the proximity of privies to their wells.

Finally, when this subject of the disposal of sewage is entertained in a proper manner, and such disposal is operated under sanitary requirements, the wonder will be how and why it was that any of the now existing methods, including the very restricted operation of cremation, for populous inland towns and cities, were tolerated.

W. H. Booth, M. Am. Soc. C. E. (by letter).—The sole reason why cremation of garbage has been a success in England is that it is burned in ovens and not near boilers; that is to say, the cells in which it is destroyed are wholly of firebrick, and only when combustion is complete are the products brought into contact with the comparatively cold plates of a steam boiler. Mr. Booth.

There is really no very serious difficulty in completely destroying refuse at a high temperature, as a simple calculation will show. Anyone who has a garden knows that ordinary garden rubbish, leaves, etc., will destroy itself if heaped up high above a small quantity of dry stuff. The initially dry stuff dries out the wet stuff which lies upon it, and this then burns and the whole heap is dried off and burned.

In England all refuse is collected together, *i. e.*, not separately, and includes ordinary house refuse, such as cinders and vegetable refuse. It is considered that this material contains 30% of water, 30% of combustible matter and 40% of inert matter, such as gravel, broken crockery and clinkers or ashes.

As carbon contains 14 647 B. T. U. per pound, 1 lb. of average refuse will contain 4 400 B. T. U. At 1 500° Fahr., the water will carry off less than 550 B. T. U., including latent heat and superheat. There will remain 3 950 B. T. U., to deal with. Allowing 18 lbs. of air per pound of fuel, there will be 54 lbs. of air, 0.3 lb. of fuel and 0.4 lb. of mineral matter to be heated up to a temperature of, say, 1 500°, or a total of 6.1 lbs. to share in 3 850 B. T. U., or 631 units per pound. The mean specific heat of all furnace products and clinker, etc., is less than 0.25. Calling it 0.25, the resultant temperature would be 2 524° Fahr., or 631 × 4, which is sufficiently above 1 500° to allow for further excess of air. Moreover, as the air forms by far the larger weight of stuff passing through the furnace, it becomes a simple matter to add considerably to the temperature by heating the air in the flues beyond the destructor.

The reduction of temperature, due to the increase of specific heat

Mr. Booth. of gases at high temperatures, need not be discussed. It will not pull down the temperature below 1 500 degrees.

The combustible elements in any refuse can be ascertained easily by having a bulk sample weighed and burned off in a gas retort, after being first dried, and again weighed, the second difference being combustible. If a refuse is so poor that it has not sufficient fuel in it to burn thoroughly, it ought to be mixed with the cheapest grade of coal obtainable and all burned together in a refractory cell. The placing of a boiler as one boundary of a destructor cell is the worst possible practice. Even good bituminous coal cannot be destroyed by such a device without smell and fumes.

The sloping grate, of course, is a great aid to the work, for the labor then consists only in helping forward the clinkered purified residue. The heat carried off by the water is partly recovered when the gases are passed finally through a boiler; but no sanitary engineer now attempts to use the heat from a destructor until destruction is complete and nothing but glowing gas remains. A simple calculation for any ordinary sample of refuse will show that it has power to heat up everything it contains to a clear red heat or higher.

Mr. Young. H. A. Young, Jun. Am. Soc. C. E. (by letter).—Mr. Parsons has made the statement that "an incinerator, * * * which has to have a fume or stench cremator, is false in design," etc. The writer believes this statement to be entirely correct. Such a crematory should be condemned at the outset. If proper combustion is obtained, no fume cremator will be required.

In the best forms of the English crematories, there is always a combustion chamber. The gases evolved by the burning garbage pass first into this chamber after leaving the furnace. Here fresh air is supplied, furnishing oxygen for the almost complete consumption of the obnoxious gases. The temperature in this chamber rises higher than that in the furnace proper.

From the combustion chamber the hot air and gases, resulting from a practically complete combustion, are carried under the inclined drying plane, thus drying the "green" garbage from the bottom to the top of the pile, and not merely the upper layer. From here the flues continue under the boilers and to the chimney.

In many of the best types the flues pass through an additional chamber, in which is installed a blower, driven by power furnished by the boilers. The draught is thus induced, and there is no danger of forcing burning particles out of the chimney.

Mr. Hering. Rudolph Hering, M. Am. Soc. C. E. (by letter.)—The discussion has revealed a fair unanimity of opinion regarding the fact that the garbage question has now reached a point where its proper solution requires the aid of the engineering profession, instead of being left for solution almost wholly to committees of city councils or of boards of health. It is also apparently agreed that different conditions demand different methods of disposal, and that each municipality should select the method which, after a careful study of the specific local conditions and a careful investigation of the cost of securing unobjectionable results, proves to be most economical. Mr. Hering.

It is further agreed that the problem must include a consideration of the methods and cost of handling all of the several kinds of refuse between the points of origin on the streets and the points of final disposal.

When opening the subject, its largeness precluded entering into details, but a number of these have been elaborated subsequently, and have added valuable information.

In reply to Mr. Baldwin the writer may state that there are at least three garbage furnaces in America, where, it is said, all the refuse is burned without the regular addition of coal, excepting to start the fires on Monday mornings, or whenever they have been interrupted. One is in Montreal, where the garbage is mixed with dry refuse and with domestic ashes, which, however, are screened at some seasons, if not always. Another is in San Francisco, where there is a similar furnace and where the garbage is likewise mixed with dry rubbish and ashes. The third is in Atlanta, Ga., where the garbage and dry rubbish alone are burned, and no domestic ashes, with the residuum of unburned coal, are added.

The writer was glad to hear from Mr. Parsons that his investigation had convinced him that the garbage and refuse in New York City, when mixed, would be self-combustible. Steam ashes, being the result of intelligent and economical firing, have no fuel value left, and always should be removed separately and utilized for filling, road-making, or as the body-material of mortar. But domestic ashes, containing 20% of unburned coal, have a fuel value beyond the cost of screening out the fine ashes, which value may be sufficiently great to utilize. Whether or not such utilization is economical, depends on the extra length of haul to the crematory, and on the cost of screening.

It would be very interesting and useful if the precise facts leading up to Mr. Parsons' conclusion could be made public. Generally, he intimates, the scientific treatment of garbage is fairly understood and could be put into practice, provided the means were available. But this understanding is not the only requisite for a solution of the present difficulties. It is also necessary that the public should appreciate it fully and have confidence in the recommendations, because the public must furnish the means to carry them out. To this end, an actual demonstration on a practical scale seems to be the *sine qua non*, showing that the treatment which engineers, after studying the special problem, believe to be the proper one, will be both successful and economical. Such a demonstration is yet lacking in our own country.

Most of the present disposal works were proposed by promoters, who succeeded in obtaining contracts to erect plants, and whose

Mr. Hering. interest lies chiefly in immediate profit to themselves. An impartial and thorough investigation, in order to ascertain the best methods and results for ultimate instead of for immediate profit has not yet been undertaken by any city in this country. Dr. Soper states that the present moment is propitious for such an investigation to be made as regards New York conditions, where the existing methods are yet crude and quite expensive, even though 40% of the work of collecting and disposing of the garbage is done by private parties, over which the city has but little control.

The writer suggests that in his opinion a further profitable investigation would be a full and detailed report on the entire subject, containing itemized expenses, and setting forth clearly the methods of operation and the results obtained in those cities of Europe where this subject has been settled satisfactorily, and, for comparison, an equally detailed report on the existing conditions in several of our own cities. Such a comparative report, the writer believes, would be of great value in pointing out specifically and most forcibly the courses to be pursued in solving this problem at home. In fact, unless we either present the public with actual information of this kind or make an actual impartial demonstration in one of our large cities, it would seem that the proper solution of the garbage question in this country will be still further postponed.

Mr. Ball makes an appeal for the disposal of street sweepings by burning instead of by filling in land, and offers many statements to support it. Such a disposition is made in several cities of Europe. In view of the fact that this material forms a rich field for disease germs, its incineration is certainly to be preferred when practicable. There are, however, many cases, particularly in our country, where the quota of combustible matter is reduced so greatly by admixture with mineral matter from badly paved or unpaved streets, that burning becomes expensive. In such cases filling in land is not only cheaper, but may be without sanitary danger. Mr. Ball further discusses the demolition of infected buildings, and gives the precautions which should be taken in their demolition. This subject had not before been touched upon, and is of sufficient importance to have been made a part of the subject under discussion. The writer believes Mr. Ball's precautions to be wise and practicable.

Commissioner Woodbury, by his remarks, leaves a doubt resting on the financial preference of the incineration over the reduction process, yet produces no facts to substantiate it. Mr. Parsons, on the other hand, unhesitatingly states his belief that incineration of the garbage of New York City is cheaper than its reduction, but, likewise, without substantiation. These, and other statements, confirm the opinion expressed by the writer, that experiments on a large scale, combined with an exhaustive inquiry into the subject, are required to demonstrate the best as well as the most economical disposal.

Rudolph Hering (1847-1923)

Rudolph Hering, called the Dean of Sanitary Engineering, was born in Philadelphia in 1847. Although American by birth, he was educated in Germany and was graduated as a civil engineer from the royal Saxon Polytechnic Institute in 1867.

His early career included land surveying for Prospect Park in Brooklyn, and being an astronomical observer in the Federal government exploration of Yellowstone Park in 1872. In the following three years he designed and supervised the construction of the Gerard Avenue Bridge in Philadelphia; it was the largest high level bridge in America at the time. His next assignment changed the course of his career. He was appointed Assistant City Engineer of Philadelphia in charge of both bridges and sewers. His reputation was such that the National Board of Health selected him to make a detailed investigation of European sanitary engineering practice. His report was a landmark state-of-the-art paper that had an important influence on American engineers. Following the publication of his report Hering established himself as a private consultant. His early projects included a water supply scheme for Philadelphia and his design for the Chicago Drainage Canal. For the next thirty years he was in consulting practice in New York City.

Hering was a leader in recognizing the significance of biochemistry and the germ theory to the treatment of water. He was the first American engineer to recommend mechnical filters and to use them for the Atlanta water supply system. He introduced the Imhoff tank in America in 1910. During his career he served as a consultant to more than 150 cities, in North and South America. He faithfully served the ASCE as a member, director and vice president. The Society established the Rudolph Hering Medal in his honor for the best contribution to sanitary engineering. In 1907 he received an honorary degree from the University of Pennsylvania and a year before his death in 1923, his alma mater, the Polytechnic Institute of Dresden, awarded him an honorary degree of Doctor of Engineering.

SELECTED BIBLIOGRAPHY
ON
HISTORY OF WATER SUPPLY, ENVIRONMENTAL
ENGINEERING AND HYDRAULICS

1. A Symposium, Historic Review of the Development of Sanitary Engineering in the United States, During the Past One Hundred and Fifty Years, Transactions ASCE, Vol 92, 1928.

2. Baker, Moses N. The Quest for Pure Water, American Waterworks, 1948.

3. Blake, Nelson Manfred. Water for the Cities. A History of The Urban Water Supply Problem in the United States. Syracuse University Press, 1956.

4. Burr, William H.; Hering, Rudolph and Freeman, John R. Report to Commissioner of Water Supply, Gas and Electricity, Commission on Additional Water Supply, N.Y. City, 1903.

5. Chesbrough, E.S. Chicago Sewerage: Report of the Results of Examination Made in Relation to Sewerage in Several European Cities in the Winter of 1856-1857. (no other information)

6. Draffin, Jasper O. The Story of Man's Quest for Water, Garrard Press, Champaign, IL, 1939.

7. Ewbank, Thomas. A Descriptive and Historical Account of Hydraulic and Other Machines for Raising Water, Ancient and Modern, Charles Scribner, New York City, 1870.

8. Hazen, Allen. The Filtration of Public Water Supplies, John Wiley and Sons, Inc. New York, 1895.

9. Herschel, Clemens. Frontinus and the Water Supply of Rome. Longmans, Green, London, 1913.

10. Maas, A., Muddy Waters, The Army Engineers and the Nation's Rivers, Harvard University Press, Cambridge, Mass., 1951.

11. Mathews, William. Hydraulia: An Historical and Descriptive Account of the Water Works of London and the Contrivances for Supplying Other Great Cities, in Different Ages and Countries, Simpkin Marshall & Co., London 1835.

12. Morgan, Arthur E. The Miami Conservancy District, McGraw-Hill, New York, 1951.

13. Robbins, Frederick W. The Story of Water Supply, Oxford University London 1946.

14. Rouse, Hunter and Ince, Simon. History of Hydraulics, Dover Publications, New York, 1963.

15. Rouse, Hunter. Hydraulics in the United States 1776-1976, Iowa Institute of Hydraulic Research, Iowa City, Iowa, 1976.

16. Smith, Norman. A History of Dams, Citadel Press, Cecaucus, N.J. 1972.

17. Van Deman, Esther B. The Building of Roman Aqueducts, Carnegie Institution of Washington, Washington, 1934.

18. Walker, M.E.M. Pioneers of Public Health, Oliver and Boyd, London, 1930.

19. Wegman, Edward. The Water Supply of the City of New York, 1658-1895, John Wiley & Sons, New York, 1896.

20. Freeman, John R. Report on New York's Water Supply, M.B. Brown Company, 1900.

21. Black, William M. Report on the Location of Sewer Outlets and the Discharge of Sewage into NY Harbor, M.B. Brown Company, 1911.

22. N.Y. Interstate Commission (NY-NJ-CT). Annual Reports, 1937-to date.

SUBJECT INDEX

Page numbers refer to first page of paper

AUTHOR INDEX

Page number refers to first page of paper